I0493974

Best of H+ Magazine

Volume 1: 2008 – 2010

Edited by R.U. Sirius
with Ben Goertzel and David Orban

http://hplusmagazine.com

http://humanityplus.org

Copyright © 2013 Humanity+
All rights reserved.

ISBN:1496073312
ISBN-13:9781496073310

Table of Contents

Preface by R.U. Sirius

Evidence Leads One to Suspect the Plausibility of the Coming of Humanity+

Strange days have found us. While we dread over oil spills, national debts, water shortages and *Jersey Shore,* Craig Venter announces the dawn of synthetic life. The manipulation of molecules gets ever more precise. The tools that bring us into the virtual world of human minds and their creations get pocket-sized... fingernail sized... and start to crawl under our skin. Athletes with replacement limbs are accused of having an unfair advantage. Bots drive cars. Optogenetics promises to stimulate brain activity without surgery. Companies race to re-engineer the brain.

The Singularity may – or may not – be near (and it may or may not *be* a Singularity), but the Science Faction future is definitely *now.*

Since 2008, *H+* Magazine has examined the present and the future largely in terms of how technology might enhance humanity (Humanity+) and allow us to self-direct our evolution. While there are no guarantees we won't screw it up, pieces are coming into place that make a different kind of future possible — a future without scarcity, without extreme psychological suffering, without aging, with very few limits on expressing and realizing the contents of human imagination (for good or for ill), and, perhaps, without death.

With sections exploring citizen scientists, AI, biotech, nanotech, longevity, human enhancement, and fun — and with an arts and culture section digging into the ups and downs of life on an ever-accelerating edge — you should consider this book a presentation of the evidence for Humanity+. So pass this on to your friends and neighbors, virtual and real; to your relations and colleagues; to anyone literate and thoughtful, willing to look at the possibilities.

These are strange days, indeed. Times are tough. Keith Richards took all the drugs and Steve Jobs took all the money. But it ain't over 'til the fat lady sings... *At your 160th birthday party.*

Preface by Ben Goertzel

These are strange days, as RU says – but also fascinating and thrilling ones!

It is, in fact, our great privilege to live in an era when H+ – the extension of humanity beyond its traditional biological form – is not merely fantasy but a reasonable description of the practical, everyday unfolding of science and technology.

H+ Magazine has provided a venue for edgy, creative thinking about H+ technologies and ideas since it was founded in 2008.

This volume collects some of the best H+ Magazine articles from the magazine's first few years, when it was edited by legendary futurist R.U. Sirius.

In these amazing, fast-changing times, even the stolid old "mainstream media" is full of articles about advanced technology and its implications. But venues like H+ Magazine can go further out on interesting limbs than Wired, Forbes and the like – as the articles in this volume amply demonstrate!

I've been involved with H+ Magazine since its creation by James Clement, Dan Stoicescu and R.U. Sirius in 2008 – and it's been a fairly wild ride. H+ Magazine has evolved considerably over the years. Ownership of the magazine passed from the founding nonprofit Humanity+, to the for-profit company BetterHumans, and then back to Humanity+. There was a paper issue that appeared in bookstores in Fall 2009; but after this sole experiment the magazine went back to its previous, purely online format. In the early years the magazine had both a standard Website format and a highly graphically slick PDF format; later the presentation was streamlined and only the standard Web format retained. But through these changes, one thing remained constant: the dedication to no-holds-barred, creative thinking about the future of humanity and transhumanity.

Each article given here, is provided along with the URL of the online version, as of 2012. In the unlikely event that these URLs change, the online versions will surely be findable via search engines. The online versions, in some cases, contain figures or videos that are omitted here. For this book version we chose to focus on the text. Multimedia online information is fantastic, but there's also something to be said for the absorption of ideas from text in book format. Strength in diversity!

I admired R.U. Sirius's writing in *Mondo 2000*, back when I was a young math professor at the end of the previous century. So it was a great pleasure for me to work with RU on H+ Magazine, for a while in a dual role as his boss (due to my role on the Humanity+ board) and employee (due to my role as occasional writer for the magazine, which he was editing). He was a delight in both roles – just as I expected, he approached the work with a delicate sensibility for transhumanist thinking and a fantastic sense of humor.

RU not only edited the magazine during the period when the articles collected here were published, but also performed the initial selection of which articles to include in this collection. So the bulk of the editorial credit for this volume belongs to the inestimable RU, although it was David Orban and I

who ultimately followed through with the preparation of the manuscript for publication, and tweaked the article selection slightly.

For a few years I used, as my email signature, Friedrich Nietzsche's quote "My humanity is a constant self-overcoming." The message RU consistently emanates, in his writing and his editing, is that this process of constant self-overcoming can often be quite a blast.

Onward and upward -- read and enjoy...!

The Rise of the Citizen Scientist

Citizen-Scientist Joseph Jackson
and the New Open Source

http://hplusmagazine.com/articles/science/citiz
en-scientist-joseph-jackson-and-new-open-source

Alex Lightman

As I wrote this, we were a few weeks away from the *H+ Summit: Rise of the Citizen-Scientist*, held at Harvard University Science Hall, June 12-13, 2010, Cambridge, Massachusetts (just north of Boston). I chaired the event, which is organized with the Harvard Future Society.

One of the speakers was Joseph Jackson, a Harvard graduate and an emerging leader in the open science movement. Part of the fun of spending time with smart people is being able to give them credit for life-changing ideas and insights. Joseph Jackson was the man who set my mind on fire with the concept of the Citizen-Scientist. As soon as Joseph explained what he meant by Citizen-Scientist, I couldn't stop thinking about how important it was to have a conference that combined H+, advanced technology, transhumanism, and scientific movements that could shift science from an occupation of hundreds of thousands of practitioners into one with hundreds of millions, all equipped with state of the art yet inexpensive equipment. Here's Joseph Jackson, in his own words, about why Citizen-Science matters.

H+: What is a Citizen-Scientist?

Joseph Jackson: A Citizen-Scientist is anyone who uses the scientific method to investigate themselves or their environment to answer a particular question or satisfy their curiosity. Several exemplary historical citizen scientists come to mind. Thomas Jefferson is the archetype of the gentleman scholar. Benjamin Franklin invented bifocals when he got tired of switching between two pairs of glasses and of course, famously flew a kite in a lightning storm to discover the principles of electricity. Edward Jenner discovered inoculation and performed the first vaccination against smallpox. Jenner's case is especially important as it highlights the power of user innovation. As a country doctor, Jenner observed that milkmaids who interacted with cattle infected with cowpox did not contract smallpox. He then transferred pus from a milkmaid to a young boy, completely protecting him from smallpox. The medical establishment was reluctant to accept the findings of a "lowly" country physician, but eventually Jenner prevailed. Thomas Edison also partially fits the descriptor of Citizen-Scientist but, because he was, frankly, a bit of a bastard (see his feud with Tesla and other abusive monopolistic industrial practices), his example is not one we want to encourage under the new Open Science paradigm. Most importantly, in the 21st century, for the first time, the plummeting costs of technology enable anyone to be a citizen scientist, whereas the classic citizen scientists of the first enlightenment were all wealthy men who had the time and resources to conduct experiments

H+: What did you study at Harvard, and how do you feel about going back as a speaker and the person who inspired the theme?

JJ: At Harvard I studied political philosophy, or "political science," in the mistaken belief that government and politics were one of mankind's most important tools for solving collective action problems. I've since become an anarchist! In 2002-2003, I took a bioethics seminar with Michael Sandel, who was then serving on President Bush's hyper-conservative Council of

Bioethics alongside chairman and arch-Luddite, Leon Kass. I had not discovered transhumanism yet, but found myself as the lone champion of H+ values in a class of shockingly close-minded fellow students. The course culminated in a visit to Washington to observe the council, where I confronted Dr. Kass about some of his positions, to no avail. This experience prompted me to seek out transhumanism.

My time at Harvard was a profoundly unpleasant ordeal, as I was largely isolated from other technoprogressive thinkers. Things came to a head when a classmate committed suicide shortly before graduation. Such an event, while a tragedy, was by no means an anomaly. There is, on average, one a year. Already battling a serious depression, I then experienced an acute crisis of meaning. I felt that I had wasted my time and that everything I had believed in was false. How could an institution that, at the time, had a $20 billion endowment, create such a "misery factory?" What should have been a secure environment for creative people to develop and flourish, in some instances became instead a cutthroat, vicious, nasty place, often churning out stunted, warped, neurotic, withered, dysfunctional, anti-social robots. Rather than encouraging graduates to take risks, innovate, and found bold new initiatives, most students were herded into a few limited career tracks... they were going to join investment banks, consulting groups, and law firms... all institutions implicated in the mega-meltdown of 2008. While MIT and Stanford are arguably more H+ friendly and have a better culture of entrepreneurship, our higher education system, in general, must adapt to remain relevant.

At the six-year anniversary of my graduation, I am pleased to return to Harvard to celebrate H+ values and help chart a course for Citizen-Science, a participatory paradigm that speaks to the concerns of all who are currently dissatisfied with our prevailing system.

H+: What are you working on?

JJ: I currently pursue three related projects. I am CEO of Lava-Amp, a company I co-founded with Venezuelan computational biologist Guido Nunez to develop extremely low-cost DNA amplification technology. PCR, which Lava-Amp performs, is the fundamental technique of molecular biology. Our device brings the cost of the hardware down from thousands of dollars, to hundreds, enabling portable DNA testing in the field, and also providing an affordable system for garage bio-hackers and amateur biologists learning the basics.

See: http://lavaamp.wordpress.com/

I also convened the first Open Science Summit, July 29-31[st,] 2010 at Berkeley. Finally, I am co-founder of BioCurious, the first Bay Area community lab for citizen science and "biohacking." The DIY biology movement has been gaining ground for the last two years with many groups meeting informally. However, we still lack infrastructure and access to equipment that is beyond the budget of a garage hobbyist. To remedy this, and also assuage some of the concerns with garage biohacking, we've pooled resources and gathered a group of Bay Area activists to found a lab for non-institutional science. Here, members can learn essential skills in a safe setting and work on projects together, instead of toiling at home and having to ask questions online... There is a lot to be said for in-person trouble-shooting.

H+: Tell me about your open science conference?

JJ: The Open Science Summit calls for a renegotiation of the social contract for science. Please see my blog post, "Enlightenment 2.0[1]," at the conference site, for the full vision. Science is the tool of tools, the method by which humanity improves its condition. However, few are thinking holistically about how to optimize the functioning of our system. Over two

---

[1] http://opensciencesummit.com/2010/05/12/enlightenment-2-0-

and a half days, we will gather stakeholders seeking to liberate our scientific and technological commons and enable a new era of decentralized, distributed innovation to solve our greatest challenges. We'll consider the following urgent policy questions: the future of biology including DIY biotech[2]; personal genomics, synthetic biology[3] (and its attendant biosecurity concerns) ; gene patents[4] after the Myriad decision; the future of peer review and scientific reputation; open data; open access publishing; novel funding mechanisms for research (crowdsourced microfinance to harness collective intelligence and allow the public to directly support projects rather than route everything through a bureaucratic grant system and many more. The solutions we'll map combine to enable a radically more efficient, collaborative, and transparent science/innovation system so that we can have a speedy, safe, and successful Singularity.

H+: What do Citizen-Scientists have to contribute to the world?

JJ: Citizen-Scientists are the force that keeps science accountable. They bring science out of the classroom and into the community, moving discoveries from the ivory tower into everyday life, whether in the form of backyard biology (see recent examples[5] of students doing DNA barcoding to identify species) or amateur astronomy (where networks of hobbyists can map the stars as well as any PhD).

Most significantly, Citizen-Scientists put a public face on science and technology. They are our number one hope for counteracting the fear and hysteria that opponents of H+ promote. Often, the public mistrusts technology because it is something foisted on them by governments, the military, or corporations. The outcry over genetically modified organisms

---

[2] http://hplusmagazine.com/articles/bio/why-diy-bio
[3] http://hplusmagazine.com/articles/bio/adventures-synthetic-biology
[4] http://hplusmagazine.com/editors-blog/aclu-v-gene-patents-and-winner
[5] http://diybio.org/2009/12/31/dna-explorers-at-nyc-high-school/

largely stems from alarm at the prospect of a few multinational entities attaining control over our food supply. The Citizen-Science and Open Science movements reverse this relationship, "domesticating" technology and re-integrating science into our lives on a human scale.

H+: What is the connection between open science, Citizen-Science and Transhumanism?

JJ: I came to transhumanism because it attempts to articulate a philosophy of life rather than an apology for death, to overcome our arbitrary limits instead of apathetically accepting them, and above all, because it dares to proclaim the possible. Sadly, most of humanity does not share our vision. Transhumanism is often perceived negatively, and sometimes portrayed as the refuge of uncaring narcissists striving for ageless omnipotence, or as a clique of techno-elites.

Open Science is *the key* to successfully harnessing game-changing transhumanist technologies for the benefit of all. All sorts of biases and agendas creep into our science and technology policy, affecting which paths are taken, and who controls the outputs of research. The current patent system's insidious effects on biomedical innovation are instructive. In what should be an era of accelerating progress, we're stuck with crude monopoly mechanisms (20-year patent terms) that evolved in an industrial economy that failed to anticipate the networked flows and collaborative sharing that a knowledge economy requires. The extension of patents to all sorts of new subject matter in the mid-1980s, and the cancerous metastasis of laws, lawyering, and litigation throughout the innovation system, delay and distort the products that reach the market. If we don't correct the flaws in the paradigm, we'll suffer horribly as lifesaving and enhancing breakthroughs are further delayed and dangerously suboptimal technological standards are adopted.

The history of technology is replete with examples of individuals and companies abusing their position to control and constrain end-user behavior. I already mentioned how Edison feuded with

Tesla, embarking on a disinformation campaign in the AC vs. DC "War of the Currents." The early adoption of a standard, be it the QWERTY keyboard, VHS vs. Betamax, or even something as mundane seeming as the screw (standardized in the 19th century), creates path dependencies and has profound effects on future innovation in a field. We've muddled through until now, but if obsolete business models focused on proprietary short-term advantage lead to the wrong platform in synthetic biology or nanotech, it may be game over. Digital rights management (DRM) is problematic enough… Imagine "Neurological Rights Management" asserted over your brain-machine interface.

There is a way to fund R&D and commercialize technology without abusing the customer/user/citizen and restricting his or her freedom. The term Open Source evokes some of this, but I've started to think the concept of technological freedom is more apt. We must change our relationship to technology, and reaffirm the basic freedoms to tinker, innovate, and hack ourselves, our products, and our world.

H+: What should H+ be focused on?

JJ: H+ must focus on rallying its members to *act*; become a DIY biologist or neuroscientist; join a hacker space, learn to be a *maker*… Spread these ideas. Be a citizen scientist, because scientists are not going to solve our problems if left to their own devices. Like all humans, they respond only to the incentives they face. We must change those incentives before it is too late.

In the words of futurist Buckminster Fuller: "In contradistinction to the esteem in which world society now holds them, scientists are the most confused and irresponsible human beings now alive. They lay "eggs" —and the businessman sells the eggs to the politicians and the politicians "scramble" or "drop" or "easy-over" those eggs as we hurtle toward oblivion. If our lives are left to their care we will all soon be dead." *(Utopia or Oblivion).*

Why DIY Bio?

http://hplusmagazine.com/articles/bio/why-diy-bio

Andrew Hessel

One of the ideas I champion is that DNA is a programming language for living things. By stringing DNA bases together in different ways, one gets different organisms. With one sequence, a bacterium is the result. With another, a butterfly. The same can be said about any subcomponent of life, all the way down to

As we get better at "printing" DNA with automated synthesizers, it gets easier to make DNA-based programs, from simple scripts (instructing a bacterium to make a new protein or compound) to whole new operating systems (genomes). And it's just gotten easier, faster and cheaper — a biological version of Moore's Law. With DNA synthesis, metabolism can be shaped by anyone who can master various DNA design tools. It's the start of a whole new era in biology: Digital biology.

I started focusing on DNA synthesis about ten years ago. At the time, I worked for a large biopharmaceutical company. As with any language, mastering DNA means one must learn to read, comprehend, and write. We had a fantastic bioinformatics team. We bought a subscription to Celera, the company Craig Venter created to sequence the human genome. With reading and comprehension well taken care of, it made sense to start thinking about how to write DNA code better.

Celera was possible because people had spent decades improving DNA sequencing technology. Still, the state of the art of DNA synthesis was poor, with low throughput and high cost (on the order of $10 per base pair). Making even a small protein (roughly 1000 bases) was expensive, and only justifiable for things like small, high-value proteins such as a growth hormone.

But I believed that as synthesis costs fell over time, less lucrative applications or experimental designs that had a higher probability of failure would fall within reach. Moreover, the work would become increasingly computer-based, rather than being done in the laboratory. Genetic engineering would come to resemble software engineering, except the programming would be biochemical.

In 2003, I took a year off to digest past experiences and to consider where life science may be going in the near future. In the meantime, digital biology got a name: synthetic biology. A small group at MIT was leading the way with DNA modules they called BioBricks that could be snapped together like Lego blocks and then easily reconfigured. The next year, they developed a student training program with BioBricks and challenged student teams to be creative in designing and making applications. Almost overnight, the genetic engineering capability once available only to the experienced and well-financed became available to relative novices for a fraction of the price.

Around this time, I found myself thinking a lot about open source versus proprietary software. The success of open source software, like Linux and Apache server, had demonstrated that community-based development could rival the work done in dedicated companies. Was open source biology possible? I believed strongly that synthetic biology, done openly, could eventually compete with the for-profit biotechnology industry. I could see a day where almost anyone with a laptop could start to create software for cells. What would people make? The projects developed by students with BioBricks suggested a broad range, from fun (bacteria programmed to smell like bananas or wintergreen) to commercially useful (next generation biofuels like butanol).

We get better at "printing" DNA has gotten easier, faster and cheaper – a biological version of Moore's Law.

By 2005, several synthetic biology companies had appeared. They'd attracted large investments from top-tier venture groups.

The field was hot. I began to think seriously about creating a Linux-style company to make drugs. How would the company be financed? How would people work together? What would they make?

Eventually, I came to believe that drug development needed a complete reboot. In the wake of the Human Genome Project and increasing lab automation, life science data was exploding. Genomics had spawned proteomics and metabolomics, and even more "omics" were appearing on the horizon. Research was growing exponentially, but development,was still stuck on a linear path from discovery to the clinic that could take a decade and a billion dollars or more. The gulf between biological R&D, always wide compared to more traditional fields of engineering, was growing even wider.

I threw away the old model for making drugs and started from scratch. Synthetic biology allowed almost anything biological — from a single protein to an entire organism — to be developed using a tool that was costing less each day. The cost of DNA-based diagnostic tests were falling quickly, too. So what was keeping the cost of making drugs so high? I identified three factors. One was overhead: the physical infrastructure of labs and staff. The second was the cost of manufacturing: facilities to make large quantities of a new drug were often custom-designed and could cost hundreds of millions of dollars. The third was the cost of clinical trials, necessary to prove to regulators that a drug was effective and safe.

Then it hit me. What if, using synthetic biology, we made drugs for just one person at a time? Fully individualized (n=1) medicines? Done open source and virtually, the overheads would be very low. Large manufacturing plants wouldn't be necessary. Best of all, the cost and complexity of clinical trials would be reduced, potentially saving years of time and massive amounts of money. Suddenly, the idea of open source drug development didn't seem farfetched.

Cancer was the perfect target to test this idea. Because cancer results from the corruption of a person's DNA, and no two people have the same DNA, each cancer is unique. A customized drug would be the ideal drug, but wasn't economically viable — at least until synthetic biology. I needed a therapeutic agent that was flexible and could be programmed. That's when I learned about oncolytic viruses — benign viruses that can infect cancer cells and kill them without affecting normal, healthy cells.

In September 2007, I gave a short talk at Aubrey De Grey's SENS conference in Cambridge outlining my intention to found an open source biotech company that would make customized therapies for breast cancer. The response to the presentation was predictable: many had concerns whether regulators would allow such a drug to be used in a human trial. I had no idea, but I knew the only way to truly find out would be to try. It took almost two years of discussion and feeling my way around, but this company now exists.

Cooperatives are community-owned and operated enterprises that exist to serve their memberships. They are corporations that operate as non-profits and can have broad membership, because people don't need to be qualified investors to get a share. They can raise substantial sums by attracting a large membership — an army — something that is fairly easy to do these days because of social networking sites like Facebook. Members of the cooperative are united by their common interest, in Pink Army's case, better, faster, and less expensive treatments for breast cancer.

Breast cancer is the first target, but ultimately the cooperative's goal is to open a path from diagnostics to the clinic for individualized medicines — to make effective cancer treatments as fast as diagnostic data can be translated into designs, manufactured, tested in the lab, and approved for use on a single person. Using open source synthetic biology, each of these steps can be automated, and each should get cheaper over time. If it works, drug development could become a real technology.

Pink Army, then, is the first DIY drug company. It's a container that allows people interested in tackling cancer to connect and focus their passion, skills, and other resources. It takes cancer — a field that has mushroomed to become a vastly complex global R&D enterprise — and reduces it to an easy-to-understand, manageable task: finding better ways to analyze and treat just one person; ways that can connect experts and resources no matter where they are in the world, ways that are safe, and ways that can scale and become more affordable as they do.

My role in the company is to share stories and make connections, something that, as a generalist, I absolutely love to do. More people are connecting every day, and the company intelligence is growing. For Pink Army to work, it needs to resonate with many people, for many different reasons. It must somehow convey the message that although cancers can arise in countless ways, the goal for treatment is almost always the same: selectively shut down or kill the broken cells.

Why am I passionate about DIYbio and open biology? Mostly it's because I think that collectively we can do better than we have. The transistor and the structure of DNA were discovered within six years of each other. Recombinant DNA technology and the microcomputer both appeared in the early 1970's. Both became big industries, but with very different dynamics. Computers are ubiquitous, while biotechnologies remain a mystery to most people, with few applications that demonstrate the utility and potential of the field to make the world a better place.

The biotech industry has struggled economically and is reaching a point where even the largest companies are resorting to merger and consolidation for growth. It's clear that something needs to change. Open biology is that change. I believe that open biology will continue to make bioengineering more accessible. It will produce new products that people want, can afford, and trust, at a much faster pace. A more open foundation for drug development could lay a strong foundation for a thriving bioeconomy that could one day be larger than the computing industry. After all, life is the most valuable commodity of all.

Re-Engineering the Human Immune System

http://hplusmagazine.com/articles/bio/re-engineering-human-immune-system

Derya Unutmaz and Gary Marcus

Swine Flu. Spanish Flu. SARS. Almost every year, it seems, there is a new virus to watch out for. Roughly thirty thousand Americans die annually from a new flu strain — meaning roughly one flu fatality for every two victims of car accidents — and there is always the possibility that we will do battle with a much deadlier strain of flu virus, such as the one (cousin to the current swine flu) that killed 50 million people in 1918.

Currently, our bodies' responses are, almost literally, catch as can. The immune system has two major components. Innate immunity responds first, but its responses are generic, its repertoire built-in and its memory nonexistent. On its own, it would not be enough. To deal with chronic infection and to develop responses targeted to specific pathogens the body also relies on a second "acquired immune system" that regulates and amplifies the responses of the inbuilt system, but also allows the body to cope with new challenges. Much of its action turns on production of antibodies, each of which is individually tailored to the physical chemistry of a particular alien invader. In the best case, the immune system creates an antibody that is a perfect match to some potential threat, and, more than that, the acquired immune system maintains a memory of that antibody, better preparing the body for future invasions from the same pathogen. Ideally, the antibody in question will bind to — and ultimately neutralize or even kill — the potentially threatening organisms.

Alas, at least for now, the process of manufacturing potent antibodies depends heavily on chance, and a type of lymphocyte known as B cells. In principle, B cells have the capacity to recombine to form a nearly infinite variety of antibodies: roughly 65 different "V regions" in the genome can combine with roughly 25 "D regions" and 6 "J regions," which further undergo random mutations. In practice, getting the right antibody depends on getting the right combination at the right time. Which combinations emerge at any given moment in any given individual is a function of two things: the repertoire of antibody molecules a given organism has already generated, and a random interplay of combination and mutation that is much like natural selection itself — new B cells that are effective in locking onto enemy pathogens persist and spread; those that do a poor job tend to die off.

Unfortunately, there is no guarantee that this system will work. In any given individual there may be no extant antibody that is sufficiently close. If there is a hole in a given individual's repertoire, that individual may never develop an adequate antibody. Even if there is an adequate starting point, the immune system still may fail to generate a proper antibody. The most useful mutations may or may not emerge, in part because the whole system is governed by a second type of immune cell known as the T cell. The job of T cells is to recognize small fragments of viral proteins as peptides and then help the B cells produce antibodies. Like B cells, T cells also have a recombinative system, generating billions of different receptors, only a few of which will recognize a given viral antigen. In effect, two separate systems must independently identify the same pathogen in order for the whole thing to work. At its best, the system is remarkably powerful — a single exposure to a pathogen can elicit a protective antibody that lasts a lifetime; people who were exposed to Spanish flu in 1918 still retain relevant antibodies today, 91 years later. (See Resources) But the system can be hit-or-miss. That same Spanish flu claimed 50 million lives, and there is no assurance that any given person will be able to generate the antibodies they need, even if they are vaccinated.

Immunity 2.0

For now, the best way to supplement the body's own defenses is through vaccines, but vaccines are far from a panacea. Each vaccine must be prepared in advance; few vaccines provide full protection to everybody, and despite popular misconception, even fewer last a lifetime. For example, smallpox vaccinations were lifelong, but tetanus vaccines generally last 5-10 years. There is still no vaccine for HIV infection. And when it comes to bacteria like tuberculosis, current vaccines are almost entirely ineffective. What's more, the whole process is achingly indirect. Vaccines work by first stimulating B cells and T cells in order to induce production of antibodies. They don't directly produce the needed antibodies. Rather, they try (not always successfully) to get the body to generate its own antibodies. In turn, stimulation of T cells requires yet another set of cells — called dendritic cells — and the presence of a diverse set of molecules called the major histocompatibility complex, creating still further complexity in generating an immune response.

Our best hope may be to cut out the middleman. Rather than merely hoping that the vaccine will indirectly lead to the antibody an individual needs, imagine if we could genetically engineer these antibodies and make them available as needed. Call it immunity-on-demand.

At first blush, the idea might seem farfetched. But there's a good chance this system, or something like it, will actually be in place within decades. For starters, as mentioned above, every T cell and B cell expresses a unique receptor that recognizes a very small piece of a foreign structure from viruses or bacteria, such as proteins. Advances in recent genetic technology have made it possible to reprogram B cells, directly or through stem cells, to produce antibodies against parts of viral or bacterial proteins. Similarly, a new clonal army of T cells that are genetically engineered to recognize parts of a virus or bacteria would help the B cells produce potent antibodies against soft spots of these viruses and other pathogens that would otherwise neutralize or kill them.

Already scientists at Caltech, headed by Nobel laureate David Baltimore, have engineered stem cells that can be programmed into B cells, which produce potent antibodies against HIV. Meanwhile, cancer researcher Steven Rosenberg at NIH has been engineering clonal T cells capable of recognizing tumors and transferring these cells to patients with a skin cancer called melanoma. His work has shown promising results in clinical trials. Together, these results could lay the groundwork for a new future, in which relevant antibodies and T cell receptors are directly downloaded, rather than indirectly induced.

Of course, many challenges remain. The first is to be able to better understand the pathogens themselves: each has an Achilles' heel, but we've yet to find a fully systematic way of finding any given pathogen's weakness, a prerequisite for any system of immunity on demand. It will also be important to develop structural models to artificially create the antibodies and T cell receptors that can recognize these regions. Eventually, as computational power continues to grow and as our structural biology knowledge increases, we may be able to design artificial vaccines completely in silico. For now, this is more dream than reality.

The real obstacle, however, is not the creation or the manufacture of protective antibodies against pathogens, but the delivery of those antibodies or cells into the body. Currently the only way to deliver antibodies into the body is difficult and unreliable. One needs to isolate stem or immune cells (B and T cells) from each individual patient and then custom-tailor the receptors for their genetic backgrounds, a process that is far too expensive to implement on a mass scale. Stem cells, nonetheless, do offer real promise. Already it seems plausible that in the future, bioengineers could create new stem cells from your blood cells and freeze them until needed. If there were to be a deadly new virus, bioprogrammers could design the potential immune receptors and genetically engineer and introduce them into your stored stem cells, which can then be injected into your blood. Eventually it may even be possible to deliver the immune receptor genes directly into your body, where they would target the stem cells and reprogram them.

At first blush, the idea of immunity-on-demand might seem farfetched.

All this is, of course, a delicate proposition. In some ways, an overactive immune system is as much of a risk as an underactive one: more than a million people worldwide a year die from collateral damage, like septic shock after bacterial infection, and inflammations that may ultimately induce chronic illness such as heart disease and perhaps even cancer. Coping with the immune system's excesses will require advances in understanding the precise mechanisms of immune regulation. This fine-tuning of the immune response could also have the bonus effect of preventing autoimmune diseases.

We are not sure when this will all happen, but there's a good chance it will, and perhaps the only question is when. There was a great leap forward in medicine when sterilization techniques were first implemented. Here's to the hope that the fruits of information technology can underwrite a second, even bigger leap.

Self Tracking:
The Quantified Life is Worth Living

http://hplusmagazine.com/articles/health-medicine/self-tracking-quantified-life-worth-living

Alexandra Carmichael

What would you do with a complete memory of your entire life? Would you relive your first kiss? Figure out what triggered your recent migraine? Remember the name that goes with the familiar face in front of you?

In other words, wouldn't it be great to have a backup of your brain?

Gordon Bell is a walking experiment doing just this. Bell has been tracking his life in delicious detail for the past 11 years. It started at Microsoft Research, where Bell started the MyLifeBits project. His goal was to digitally record as much of his life as possible. He wore a camera, recorded his phone calls, scanned photos and letters, documented all of his computer work, and tracked his biometrics. The job of Bell's colleague Jim Gemmell was to build software to make all this tracking easier, searchable, and meaningful.

This September Bell and Gemmell released a book called *Total Recall: How the E-Memory Revolution Will Change Everything.* In it, they talk about the future implications of being able to remember everything about your life in extraordinary detail. Bell proposes that a "continuous digital diary or e-memory" that integrates digital recording devices, memory storage and search engines will fundamentally "change what it means to be human." Their work includes research into memory, work, health, learning, and immortality. A side order of privacy is served up too, as the authors distinguish between "life loggers," who keep their records to themselves, and "life bloggers," who broadcast their data.

Of course, self-tracking is not a new idea. People have been recording their lives in analog format ever since they started drawing on cave walls. Benjamin Franklin used to keep a detailed checklist of the thirteen virtues he was striving to live by, including annotated explanations of where he was succeeding and where he still needed to improve.

Now, it can all be monitored digitally.

It probably won't surprise the readers of this article that I track myself. But it might surprise you that I track 40 different things every day. On a typical day, my pain level is 2, my weight is 126 lbs, I did 1 hour of walking, my happiness is 9, and I slept 6 hours. Charts like the one below help me to be aware of my mood, activity level, and sleep, and how these things interrelate.

With a background in molecular genetics and bioinformatics, as well as a history of chronic pain, I started tracking to help myself. But I soon wanted to apply what I had learned to help others. Here are two of the projects I'm currently working on.

Quantified Self

Imagine a show-and-tell for grownups. Fifty or so people get together every month in the San Francisco Bay Area and New York City. They show each other the data they've collected, the tools they've built, the ideas they have, or the self-tracking projects they're working on. Feedback and questions pop up from the audience. All of it is reminiscent of the Homebrew Computer Club.

This amazing group, which calls itself The Quantified Self, was started in 2007 by Kevin Kelly and Gary Wolf of *Wired* Magazine. They noticed a trend in people seeking greater self-knowledge, and using numbers on this quest to understand themselves. (Hence the name.)

Some of the projects that have been shown-and-told at Quantified Self meetups include:

1: Tweetwhatyoueat

Alex Rossi showed a demo of the web application he built to help people keep track of the foods they eat. He even added a crowdsourced calorie lookup, so if you're not sure how many calories were in the banana you just ate, you can see what eight other people estimate the calories of a banana to be. He used the Twitter API, with a simple prefix people can use in their tweets that will direct the information to his system. (See a video of his Quantified Self presentation in Resources below.)

2: Lifecasting

Ryan Grant showed a wearable camera he was working on that would take tens of thousands of pictures every day. That's a picture every 2 to 5 seconds. It's like a memory assistant that puts scrapbooking to shame. Of course, categorizing and searching all those photos is the next challenge. (See Resources for Ryan's talk.)

3: Fish Oil Makes You Smarter

Here is an example of pure self-experimentation. Tim Lundeen gave himself a cognitive test of 100 simple math problems, every day for 130 days. On day 80, he started taking double his normal dose of DHA (from fish oil), and his time to complete the math problems decreased. See the chart below.

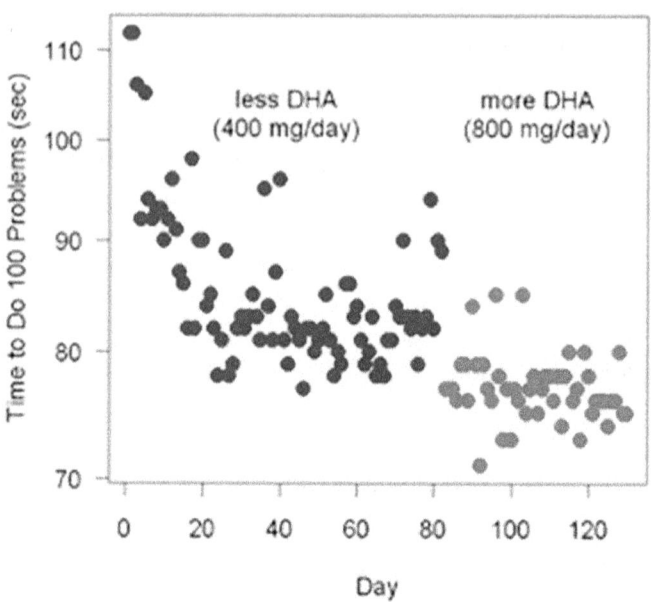

Tim Lundeen's Arithmetic Data

4: Your Genome on Twitter

At a recent Quantified Self meetup, Attila Csordas talked about his attempt to post the data from his 23andMe genome scan to Twitter, with each SNP (single nucleotide polymorphism) expressed as a tweet. What happens when more people make their personal health information public? Does the health information have a life and friends of its own? Would you follow a SNP on Twitter? These are some of the questions that arose out of this animated discussion.

5:   A Square Meal

Mimi Chun, at a New York Quantified Self meetup, showed the beautiful quantitative artwork she created based on the color of her food palette over the course of a week (see below).

(If you're reading this in black and white, see the color version at http://hplusmagazine.com/2010/02/08/self-tracking-quantified-life-worth-living/)

So whether it's for art, memory, health, or data for data's sake, people are tracking themselves and sharing their results. We do it because we love data or because we have specific things we want to optimize about ourselves. As Kevin Kelly wrote, "Unless something can be measured, it cannot be improved."

When Gordon Bell is asked what he has learned about himself through the MyLifeBits project, his reply is unexpectedly qualitative: "That's been a really hard question to answer... I guess it's the rich set of connections and people that I've been involved with."

Bell's comment reflects the challenges that come up over and over again at Quantified Self discussions — questions that tend to revolve around two topics: motivation and meaning. How do we stay motivated (and motivate others) to track ourselves, and how do we make sense and learn actionable lessons from all of this data? The search for solutions to these challenges offers ample opportunities for innovation. Imagine self-tracking games that reward people for recording their health with badges of recognition; passive monitoring devices that remove the need to actively track yourself; social pressure in the form of online group challenges; prizes awarded to algorithms that turn messy data into beautiful insight.

CureTogether

One step on this path of innovation is self-tracking applied to health. An example of this is CureTogether, a patient data-sharing site I co-founded with Daniel Reda where people come to self-report symptoms, treatments, and triggers for over 300 conditions.

People are tracking their depression, cholesterol, migraines, and countless other measures. Using migraine as an example, patients visiting CureTogether can see community statistics and learn that the top reported symptoms are "Nagging pain in one side of the head" and "nausea;" the top reported treatments are "sleep" and "ibuprofen;" the top reported triggers are "stress" and "not enough sleep" and the top related conditions are anxiety and depression.

Instead of narrative websites that provide emotional support in the form of shared disease stories, the quantitative data at CureTogether enables decision support and hypothesis generation. People are getting ideas for new treatments that they ask their doctors about. They are seeing how common or rare their symptoms are, and learning what triggers might be affecting them. While each individual's data is completely private, the aggregate data is open for researchers around the world to analyze and use to make discoveries for the greater good. Some interesting correlations are already starting to emerge, like a potential link between migraine and fibromyalgia.

Self-Tracking Will Change the Future of Health

The Quantified Self and CureTogether are just the beginning. Here are some scenarios that point to a fundamental shift in healthcare coming in the near future.

1. **Self-Organized Clinical Trials**

   Patients have started coming together to define their own case-control studies. At PatientsLikeMe, patients with ALS either took lithium or didn't take lithium, and they tracked their progress. They didn't find that lithium helped slow the disease progression, but they did run an ALS trial with the largest population in the fastest time and with the lowest cost ever.

2. **Streaming, Ubiquitous Biosensors**

   Think constantly uploading data about your body to an online repository is far off in the future? Not so. For a one-time fee of $99, you can now have FitBit, the accelerometer with the beautiful clip-on form factor and wireless uploading of exercise and sleep data. It's passive motion tracking in your pocket.

3. **Analytics for Your Health**

   A number of emerging companies are trying to do for health what Google Analytics has done for website management and what Mint has done for finances. DailyBurn is one example doing this for fitness and nutrition, with a $0.99 iPhone app that lets you take pictures of the barcodes on foods you eat to help you more smoothly track your caloric intake. A big challenge here is the lack of interoperability and standards adoption. EMRs, PHRs, and self-reported data just don't talk to each other very well yet, but medical informatics groups like the Regenstrief Institute are working on it.

4. What Treatment Will Work For *Me*?

The true promise of all this self-tracking is, in the end, personalized medicine. With enough data about your symptoms, biomarkers, environment, genes, response to previous treatments, and aggregate population data for comparison, it should be possible for a series of algorithms to determine which treatment is statistically most likely to work for you, with the greatest efficacy and least side effects.

This is an exciting future and I am dedicating all my waking effort to it. So now that you've heard Gordon Bell's story, and mine, and the voices of Quantified Selfers across the country, the choice is yours: will you document your life?

From Hackerspace to Your Garage

http://hplusmagazine.com/articles/toys-tools/hackerspace-your-garage-downloading-diy-hardware-over-web

Bryan Bishop & Surfdaddy Orca

What does open source software have in common with manufactured parts? Both involve a growing movement of "do it yourself" (DIY) packaging and distribution that follows the open source model of the free Linux operating system (OS). In other words, as with free software, you will likely soon be able to download hardware from the web in the form of free packages of coded instructions to make... Well, just about anything — from a Lego block to the jet engine of an F-16.

The impetus behind downloading DIY hardware from the web is to leverage hackerspaces (those collaborative dens of real world hacking), fab labs (fabrication laboratories), and other groups around the world to shift the burden away from "makers" (manufacturers) to a core group of package maintainers who verify hardware designs and make it easy for makers to do what they love best — making things.

And this involves making real stuff, not just code. "Recently a guy reinvented the fabric of industrial society in his garage," writes Kevin Kelly of the late Dave Gingery, a midnight machinist in Springfield, Missouri. Gingery enjoyed the challenge of "making something from nothing," and had a knack for piecing together a complete machine shop from alley scraps. "He made rough tools that made better tools, which then made tools good enough to make real stuff," Kelly continues. The spirit of Gingery lives on in the hearts of makers today.

Gingery is an interesting example of one side of DIY manufacturing equation — the fabrication of "real stuff" from machine tools such as lathes, shapers, planers, and hexapods. One example of the other side of this equation is represented by something known as the Enhanced Machine Controller or EMC. The EMC is a powerful and free software system for controlling machine tools that accepts programs in G-code (the RS-274 standard machine tool programming language). The computer code causes the machine tools to move. And moving tools make stuff.

Let's take the simple example of a Lego block. In addition to the dimensions, the object's specification includes the spacing of the stud centers, the stud diameter, and the thickness of the brick walls.

Here's an example of the encoded description of a single Lego block:

```
author:   'ben   lipkowitz' license:   'GPL2+'
urls:

http://heybryan.org/mediawiki/index.php/Skd
b -

http://fennetic.net/git/gitweb.cgi?p=skdb.g
it;a=blob_plain;f=screw.yaml'

parts: - !lego
name: technic pin
ldraw: 3673
material: ABS
files: - "gear_pin1.stp"
interfaces: - !lego_feature
point: [0.0, 0.0, 0.0]
type: pin
x_vec: [1.0, 0.0, 0.0]
y_vec: [0.0, 1.0, 0.0] - !lego_feature
point: [0.0, 0.0, -16.0]
type: pin
```

```
x_vec: [1.0, 0.0, 0.0]
y_vec: [0.0, -1.0, 0.0]
```

This hardware specification applies the lessons learned from dependency resolution in open source software development (making sure one piece of software works with another) to manufacturing and the art of building things. The goal is to download the Lego block — or any other part or assembly — over the web. This means providing instructions to a home PC to fabricate the block in your personal fab lab, or even your garage. Today people read and carry out these instructions, but computers and machines are starting to pick up the load.

The dependency tree for a Lego block includes options like plastic injection molding and 3D printing. This "metadata" information ultimately provides the instructions to build a DIY object. These instructions tend to be much more precise, machine readable, and user friendly than an "Instructable" (a set of instructions provided at the DIY website instructables.com).

Reengineering the delivery of physical products over the internet is also reengineering the social fabric to manage DIY manufacturing. The Lego block example comes from the Social Engineering-Knowledge Database (SKDB) project that provides the means for specifying dependencies and applying standards that can be shared by open source DIY makers. You don't need to reinvent the wheel every time you begin a new project. Someone may have already done most or all of the work for whatever you are trying to do, and then released the plans on the internet. And there are many common tools and parts involved in making things. The SKDB project directly tackles the challenge of packaging and distributing these plans.

The SKDB project simplifies the process of searching for free hardware designs, comparing part compatibility, building lists of materials and components, and determining where to get them by organizing them into packages. For example, in the world of the open source and free Debian distribution of the Linux OS, packages involve the apt-get command, a powerful tool that installs new software packages and upgrades existing software packages. Debian comes with over 25,000 free packages — precompiled software bundled up in a nice format for easy downloading and installation.

The open source Debian packaging and distribution model is built on the Debian social contract. This contract, familiar to open source developers, is a set of commitments that hackers — who freely contribute their time to building the OS — agree to abide by in order to use the software. Here are the basic principles:

1. Debian will remain 100% free.

2. We will give back to the free software community.

3. We will not hide problems.

4. Our priorities are our users and free software.

5. Works that do not meet our free software standards are not part of the Debian system.

Software package managers are critical to the success of Debian and Ubuntu (another free OS that proudly proclaims "you can download, use and share Ubuntu with your friends, family, school or business for absolutely nothing"). They are essential to DIY open hardware as well. Rather than spending hours trying to track down free copies of the dependencies, or searching the web for days for the right versions, package managers do the work.

And why should a hacker stop at version control alone? Why not use hardware packages to make a lab or to make tools for an experiment? Instead of painfully picking through websites like instructables.com, thingiverse.com, or hardcopy magazines, package managers do the gritty work — generate the instructions or order parts over the web — and the user is freed to make stuff, rather than chasing down dependencies. Slowly but surely, computer algorithms are taking over these tasks.

The SKDB project is like apt-get[6], but for real stuff. In the SKDB project, hardware specifications are organized into packages. Packages are a standard and consistent way for programs to find data.they can contain the following:

• CAD files

• CAM parameters

• computer-readable descriptions of product specifications,

• product-specific code,

• instructions for assembly and construction, and

• a bill of materials.

For each part in a package, there are a number of interface definitions that describe how the part can connect with other parts, even parts from other packages. Each package also lists dependencies that have to be bought or built in order to successfully carry out a project. For example, a drill press is required to make holes with a certain level of accuracy. The SKDB project downloads all of the dependencies automatically and compares them to your existing inventory, and generates instructions for your computer numerical controlled (CNC) machinery, if you have any. A CNC machine is a machine tool that uses programs to automatically execute a series of machining operations.

With OpenCASCADE, an open source CAD geometry kernel, parts can be visualized and combined in real time to show new assemblies and constructions. 3D pictures contain information that can be used to automatically generate instructions for assembling the parts and projects into human-readable and robot-readable G-code instructions. A wiki-like frontend to the SKDB project is in the process of being integrated with the free git [http://git-scm.com/] distributed revision control system so that hackers working on a project can publish and share their

---

[6] http://hplusmagazine.com/articles/toys-tools/hackerspace-your-garage-downloading-diy-hardware-over-web

modifications with the rest of the world. These tools are vital to the future success of DIY collaborative and free manufacturing. Without a solid base for sharing and building upon each other's work, the movement will flounder.

Kevin Kelly, who saw the potential to re-invent the fabric of industrial society in Dave Gingery's home fab lab, has also written some other thoughtful pieces on civilization as a sort of organism or creature or monster. "I've been thinking of civilization (the technium) as a life form, as a self-replicating structure," writes Kelly. "I began to wonder what is the smallest seed into which you could reduce the 'genes' of civilization, and have it unfold again, sufficient that it could also make another seed again. That is, what is the smallest seed of the technium that is viable? It must be a seed able to grow to reproduction age and express itself as a full-fledge civilization and have offspring itself — another replicating seed."

Kelly saw this "seed" as a library full of knowledge and perhaps tools. It was this concept of a seed library of knowledge in the form of the SKDB project — paired with Gingery machine enthusiasts pouring molten metal into the shape of metal-cutting lathes — that inspired the vision of controlling and maintaining a repository for downloading machine-readable instructions that can be converted into stuff.

Translating bits and bytes on the web into something physical (atoms) requires a mechanism. In the case of DNA, this mechanism is a ribosome that assembles proteins from the coded data on a molecule of messenger RNA (mRNA). The SKDB project can be just such a mechanism for sharing hardware over the internet — ultimately translating the contents of packages into things. This can include DIY biology lab equipment, like pipettes, test tubes, racks, refrigerators, thermocyclers, cameras, spectrophotometers, and autoclaves to work with the reusable genetic building blocks of synthetic biology.

How soon will the designs of hackerspace be made available over the internet to the machine shop in your garage? Perhaps much sooner than you would guess. Machine-readable instructions such as G-code exist today, but they aren't always human readable. Debian has an estimated $13 billion of volunteer hacker time invested in it. While the SKDB project is still in its infancy, technical demonstrations exist today and the movement is growing. "Making something from nothing" may be as close as the next hacker.

Interested in hacking open source hardware packages? Bryan Bishop and Ben Lipkowitz are forming a transhuman technology co-op powered by open source hardware. On the off-chance that you're not involved, you should be — contact the community today at:
openmanufacturing@googlegroups.com

Scrapheap Transhumanism DIY RFID

http://hplusmagazine.com/2010/02/11/scrapheap-transhumanism/

Lepht Anonym

I'm sort of inured to pain by this point. Anesthetic is illegal for people like me, so we learn to live without it; I've made scalpel incisions in my hands, pushed five-millimeter diameter needles through my skin, and once used a vegetable knife to carve a cavity into the tip of my index finger. I'm an idiot, but I'm an idiot working in the name of progress: I'm Lepht Anonym, scrapheap transhumanist. I work with what I can get.

Sadly, they don't do it like that on TV. The art of improving the human is shiny and bright in the media. You see million-euro cryogenics policies and hormonal life-extension regimes that only the elite can afford. You see the hypothesis of an immortal silicon body to house your artificially enhanced mind. You could buy that too, maybe, if you sold most of your organic body and the home it lives in. But you can do something to bring it down a notch: homebrewing.

My first foray was into RFID (radio frequency identification) — following Amal Graafstra. He's famous for having his doctor implant him with a passive ID ampoule. After one visit to an outraged state GP here in Scotland ("I wouldn't do it even if I could, and I have no idea why you want to do it!"), I was fairly certain I'd been born in the wrong country for that. Here, doctors would be struck off the records for helping me. I was on my own.

Luckily, I'm far too stupid to be stopped by bureaucracy. I bought my first Swann-Morton scalpel online, scrubbed the cleanest bathroom we could get with household bleach, settled myself cross-legged over the bathtub with my spotter, and poised the blade over the Biro-ink line I'd drawn for guidance. For a few minutes, I doubted whether I'd even be able to do it — cutting yourself open is not something we're adapted to be good at. Contemplating St. Gibson, I took the plunge.

It took a few weeks to heal, and when it did, with some help from my local gurus I was able to program a cheap open-source Phidgets RFID reader to recognise the chip's hexadecimal ID. The piece of C code that did it resided on a Linux machine and ran in the background while the reader was connected, waiting for my chip to show up. In short, it could see me and print a little "hi" when it did. That's just garbage programming, too — you can see the potential if it was given to a real coder. The chip works with any homebrew RFID project: Graafstra's RFID keyboard, for instance, grants or revokes access to my XP box based on whether the user is lepht or not. You want a laptop tracking system? A door that only lets you in? A safe that won't allow keypad input if you're not next to it? All you need is an ampoule (you get five for a euro, the last time I checked), from any RFID hobby place, a cheap reader, and a touch of disregard for risks. Salvage a keyboard from your local dump and you've got a simple system for bioidentification.

RFID chips work on passive power. Readers take power from a USB to generate magnetic fields. The chips contain copper coils to convert the magnetic field back into an electric one that they can use as their power source. After the RFID op, I acquired another implant that works with EM fields, the neodymium-60 nodule pioneered by Steve Haworth.

The implants sit in various places under my skin: middle fingertips of my left hand, back of the right hand, right forearm — tiny magnets, five or six millimeters across, coated in gold and then in silicon to isolate the delicate metal from the destructive environment of your body. They're something of an investment at about thirty euros apiece, and hard to get hold of, but worth pursuing. When implanted, they become technological sensory organs.

There's an entire world of electromagnetic radiation out there, invisible to most. Our cities are saturated with it. A radio, for instance, gives off a field that's bigger than the device itself. So do power supplies and wires in the walls. The implants pick up on the fields, and because they're magnets, they fizz with gentle electricity, telling you this hard drive is currently active, that one is turned off, there's the main line in the wall. Holding a mobile phone, you can feel the signals it sends and receives. You know it's ringing before it starts to play any sounds. And when you answer it, you stick the touchscreen stylus to the back of your hand to hold it, then to your finger to type.

After a while, you don't notice anything novel about this at all. Building computers, you pick up screws that have fallen down into the motherboard with one fingertip and stick them on the back of your wrist for safekeeping. You know not to touch the board when it's powered, because your hands can "see" whether it is or not, just like you can see whether the hard drives being tested on the machine next to it are actually being written to or not. It's just like any other sense, except that this one can be given to you for the price of a node, a needle and a bottle of antiseptic. A new way of seeing the world, all for about fifty euros. There's nothing stopping you except your own sense of self preservation. I say all this not to show off, but to invite more people in. I dream of seeing more body tweakers around who are into these things. I know there are people out there who could open up home modification like we've never dreamed.

Watching commercials for vitamin pills on TV and thinking you need a mad scientist's lab to be a transhumanist? You don't. I've got no money, talent or backing. You just need curiosity and the willingness to withstand some pain. Risk, not money, is our obstacle. Is it yours? Are you reading this magazine right now? Do you think like that? What could we achieve together?

Turn off the TV. Pick up the needle. Come to the junkyard.

Radically Enhanced Human Body

http://hplusmagazine.com/2010/11/30/top-5-human-enhancement-must-haves/

Natasha Vita-More

The World Health Organization estimates that 25 million people worldwide are affected by over 1,000 genetic conditions. There are approximately 24.6 million people alive today that have been diagnosed with cancer within the last five years. In the United States alone, 101,000 people are currently waiting for an organ transplant, and the number grows 300 people each month, according the Mayo Clinic. Over 150,000 leg prosthetics are sold annually, marking a distinct trend toward body replacement parts.

Observably, human enhancement has arrived, yet human intelligence has not advanced at the speed of accelerating technologies.What can we do to enhance cognition and also bring about a more hybrid human physiology—one that is mutable and longer lasting.

The following is a list of possible transhuman must haves for the 21st century:

1. **Brain Enhancement:** Metabrain prosthetic, which includes
   - Perception and behavior feedback app
   - AGI decision assistant
   - Cognitive error correction task pane with auto-correct options
   - Multiple viewpoints window with drop down elements
   - Neural connectome observation display

2. **Body Enhancement:** Whole-Body prosthetics, which includes
   - In vivo fiber optic spine
   - Atmospheric physical stimuli sensors
   - Solar protective nanoskin
   - Nanorobot regeneration system for cells and organs
   - Exoskeleton mobility device and 3-D printable enhancement appendages
   - Replaceable genes
   - Sex and gender modification enabler

3. **Behavior** Enhancement: Psychology prosthetic, which includes
   - Awareness Customization
   - Connectivity macros and quantified self mapping
   - Empathic ping layers
   - Finessed emotions helper
   - Persona multiplier and recorder
   - Seamless relay between platforms, avatars and syn-bios

4. **Style Enhancement:** Aesthetics prosthetic, which includes
   - Wearable or in-system options
   - Transhuman haute couture clipboard
   - Radical style click-and-drag option
   - Day-to-night shape shifting
   - Customized tone, texture and hue change

5. **System Care:** Warranty prosthetic, which includes
   - Additional 24 chromosomal pairing
   - Guarantee for genetic or code mutations or defects
   - Upgradable immune system and anti-virus system

Our Machines/Ourselves: AI/Bots/The Singularity

Singularity 101 with Vernor Vinge

http://hplusmagazine.com/articles/ai/singularit
y-101-vernor-vinge

Douglas Wolens

The Singularity. Ray Kurzweil has popularized it and, by now, some of our readers no doubt drop it frequently into casual conversation and await it like others await salvation. (The second "helping?") but many more are still unfamiliar with the concept.

The contemporary notion of the Singularity got started with legendary SF writer Vernor Vinge, whose 1981 novella *True Names* pictured a society on the verge of this "event." In a 1993 essay, "The Coming Technological Singularity," Vinge made his vision clear, writing that "within thirty years, we will have the technological means to create superhuman intelligence. Shortly after, the human era will be ended."

We caught up with Vinge at the 2008 Singularity Summit in San Jose, California, where he opened the proceedings in conversation with Bob Pisani of CNBC.

Vinge's most recent novel is *Rainbow's End*.

H+: Let's start with the basics. What is the Singularity?

VERNOR VINGE: Lots of people have definitions for the Singularity that may differ in various ways. My personal definition for the Singularity — I think that in the relatively near historical future, humans, using technology, will be able to create, or become, creatures of superhuman intelligence. I think the term Singularity is appropriate, because unlike other technological changes, it seems to me pretty evident that this change would be unintelligible to us afterwards in the same way that our present civilization is unintelligible to a goldfish.

H+: Haven't there been other Singularities throughout history?

VV: Some folks will say there have been singularities before — for instance, the printing press. But before Gutenberg, you could have explained to somebody what a printing press would be and you could have explained the consequences. Even though those consequences might not have been believed, the listener would have understood what you were saying. But you could not explain a printing press to a goldfish or a flat worm. And having the post-Singularity explained to us now is qualitatively different from explaining past breakthroughs in the same way. So all these extreme events like the invention of fire, the invention of the printing press, and the evolution of cities and agriculture are not the right analogy. The technological Singularity is more akin to the rise of humankind within the animal kingdom, or perhaps to the rise of multi-cellular life.

H+: Is the Singularity near?

VV: I'd personally be surprised if it hadn't happened by 2030. That doesn't mean that terrible things won't happen instead, but I think it is the most likely non-catastrophic event in the near future.

H+: Should we be alarmed by the Singularity?

VV: You are contemplating something that can surpass the most competitively effective feature humans have — intelligence. So it's entirely natural that there would be some real uneasiness about this. As I said, the nearest analogy in the history of the earth is probably the rise of humans within the animal kingdom. There are some things about that which might not be good for humans. On the other hand, I think this points toward something larger. Thinking about the possibility of creating or becoming something of superhuman intelligence is an example of an optimism that is so far-reaching that it forces one to look carefully at what one has wanted. In other words, humans have been striving to make their lives better for a very long time. And it is very unsettling to realize that we may be entering an era where questions like "what is the meaning of life?" will be practical engineering questions. And that *should* be unsettling. On the other hand, I think it could be kind of healthy if we look at the things we really want and look at what it would mean if we could get them. And then we could move forward from there.

H+: What signs would you look for which indicated that the Singularity is near?

VV: There are a number of negative and positive symptoms that a person can watch for. An example of a negative symptom would be if you began to notice larger and larger software debacles. In fact, that's sort of fun to write about. One of the simplest of positive signs is simply to note whether or not the effects of Moore's Law are continuing on track.

The fundamental change that may be taking place -- humans may not be best characterized as the tool-creating animal but as the only animal that has figured out how to outsource its cognition — how to spread its cognitive abilities into the outside world. We've been doing that for a little while... ten thousand years. Reading and writing is outsourcing of memory. So we have a process going on here, and you can watch to see whether it's ongoing. So, for instance, in the next ten years, if you notice more and more substitution for using fragments of human cognition in the outside world — if human occupational responsibility becomes more and more automated in areas involving judgment, that haven't yet been automated — then what you're seeing is rather like a rising tide of this cognitive outsourcing. That would actually be a very powerful symptom.

Ray Kurzweil: The H+ Interview

A 3-way conversation with the brilliant and controversial inventor and futurist

http://hplusmagazine.com/articles/ai/ray-kurzweil-h-interview

Surfdaddy Orca & R.U. Sirius

Ray Kurzweil needs little or no introduction to most *H+* readers. Principal developer of the first omni-font optical character recognition, the first print-to-speech reading machine for the blind, the first CCD flat-bed scanner, the first text-to-speech synthesizer, the first music synthesizer capable of recreating the grand piano and other orchestral instruments, and the first commercially marketed large-vocabulary speech recognition, Ray has been described as "the restless genius" by the *Wall Street Journal,* and "the ultimate thinking machine" by *Forbes Inc.* The magazine also ranked him #8 among entrepreneurs in the United States and called him the "rightful heir to Thomas Edison." His Kurzweil Technologies, Inc. is an umbrella company for at least eight separate enterprises.

Ray's writing career rivals his inventions and entrepreneurship. His seminal book, *The Singularity is Near*, presents the Singularity as an overall exponential (doubling) growth trend in technological development, "a future period during which the pace of technological change will be so rapid, its impact so deep, that human life will be irreversibly transformed." With his new films, *Transcendent Man* and *The Singularity is Near: A True Story about the Future*, hitting the film festival circuits, he is becoming an actor, screenplay writer, and director as well.

Sponsored by the Singularity Institute, the first Singularity Summit was held at Stanford University in 2006 to further understanding and discussion about the Singularity concept and the future of technological progress. Founded by Ray, Tyler Emerson, and Peter Thiel, it is a venue for leading thinkers to explore the idea of the Singularity – whether scientist, enthusiast, or skeptic. Ray also founded Singularity University in 2009 with funding from Google and NASA Ames Research Center. Singularity University offers intensive 10-week, 10-day, or 3-day programs examining sets of technologies and disciplines including future studies and forecasting; biotechnology and bioinformatics; nanotechnology; AI, robotics, and cognitive computing; and finance and entrepreneurship.

Ray headlined the recent Singularity Summit 2009 in New York City with talks on "The Ubiquity and Predictability of the Exponential Growth of Information Technology" and "Critics of the Singularity." He was able to take a little time out after the Summit for two separate interview sessions with *h+* Editor-in-Chief R.U. Sirius and Surfdaddy Orca on a variety of topics including consciousness and quantum computing, purposeful complexity, reverse engineering the brain, AI and AGI, GNR technologies and global warming, utopianism and happiness, his upcoming movies, and science fiction.

Consciousness, Quantum Computing, & Complexity

Surfdaddy Orca: I wonder if you could briefly recap some of the updates that came out of your first talk at the Singularity Summit?

Ray Kurzweil: One area I commented on was the question of a possible link between quantum computing and the brain. Do we need quantum computing to create human level AI? My conclusion is no, mainly because we don't see any quantum computing in the brain. Roger Penrose's conjecture that there was quantum computing in tubules does not seem to have been verified by any experimental evidence.

Quantum computing is a specialized form of computing where you examine in parallel every possible combination of qubits. So it's very good at certain kinds of problems, the classical one being cracking encryption codes by factoring large numbers. But the types of problems that would be vastly accelerated by quantum computing are not things that the human brain is very good at. When it comes to the kinds of problems I just mentioned, the human brain isn't even as good as classical computing. So in terms of what we *can* do with our brains, there's no indication that it involves quantum computing. Do we need quantum computing for consciousness? The only justification for that conjecture from Roger Penrose and Stuart Hameroff is that consciousness is mysterious and quantum mechanics is mysterious, so there must be a link between the two.

I get very excited about discussions about the true nature of consciousness, because I've been thinking about this issue for 50 years, going back to junior high school. And it's a very difficult subject. When some article purports to present the neurological basis of consciousness... I read it. And the articles usually start out, "Well, we think that consciousness is caused by..." You know, fill in the blank. And then it goes on with a big extensive examination of that phenomenon. And at the end of the article, I inevitably find myself thinking... Where is the link to consciousness? Where is any justification for believing that this phenomenon should cause consciousness? Why would it cause consciousness?

In his presentation, Hameroff said consciousness comes from gamma coherence, basically a certain synchrony between neurons that create gamma waves that are in a certain frequency, something like around 10 cycles per second. And the evidence is, indeed, that gamma coherence goes away with anesthesia.

Anesthesia is an interesting laboratory for consciousness because it extinguishes consciousness. However, there's a lot of other things that anesthesia also does away with. Most of the activity of the neocortex stops with anesthesia, but there's a little bit going on still in the neocortex. It brings up an interesting issue. How do we even know that we're not conscious under anesthesia? We don't remember anything, but memory is not the same thing as consciousness. Consciousness seems to be an emergent property of what goes on in the neocortex, which is where we do our thinking. And you could build a neocortex. In fact, they are being built in the Blue Brain project, and Numenta also has some neocortex models. In terms of hierarchies and number of units in the human brain, these projects are much smaller. But they certainly do interesting things. There are no tubules in there, there's no quantum computing, and there doesn't seem to be a need for it.

Another theory is the idea of purposeful complexity. If you achieve a certain level of complexity, then that is conscious. I actually like that theory the most. I wrote about that extensively in *Singularity is Near*. There have been attempts to measure complexity. You have Claude Shannon's information theory, which basically involves the smallest algorithm that can generate a string of information. But that doesn't deal with random information. Random information is not compressible, and would represent a lot of Shannon information, but it's not really purposeful complexity. So you have to factor out randomness. Then you get the concept of arbitrary lists of information. Like, say, the New York telephone book is not random. It's only compressible to a limited extent, but it's not a high level of complexity. It's largely an arbitrary list.

I describe a more meaningful concept of Purposeful Complexity in the book. I propose that there are ways of measuring purposeful complexity. In this theory, humans are more conscious than cats, but cats are conscious, but not quite as much because they're not quite as complex. A worm is conscious, but much less so. The sun is conscious. It actually has a fair amount of structure and complexity, but probably less than a cat, so...

SO: How do you go about proving something like that?

RK: Well, that's the problem. My thesis is that there's really no way to measure consciousness. There's no "Consciousness Detector" that you could imagine where the green light comes on and you can so, "OK, this one's conscious!" Or, "No, this one isn't conscious."

Even among humans, there's no clear consensus as to who's conscious and who is not. We're discovering now that people who are considered minimally conscious, or even in a vegetative state, actually have quite a bit going on in their neo-cortex and we've been able to communicate with some of them using either real-time brain scanning or other methods.

Today, nobody worries too much about causing pain and suffering to their computer software. But we will get to a point where the emotional reactions of virtual beings will be convincing, unlike the characters in the computer games today. And that will become a real issue. That's the whole thesis of my movie, *The Singularity is Near.* But it really comes down to the fact that it's not a scientific issue, which is to say there's still a role for philosophy.

Some scientists say, "Well, it's not a scientific issue, therefore it's not a real issue. Therefore consciousness is just an illusion and we should not waste time on it." But we shouldn't be too quick to throw it overboard because our whole moral system and ethical system is based on consciousness. If I cause suffering to some other conscious human, that's considered immoral and probably a crime. On the other hand, if I destroy some property, it's probably OK if it's my property. If it's your property, it's probably not okay. But that's not because I'm causing pain and suffering to the property. I'm causing pain and suffering to the owner of the property.

And there's recognition that animals are probably conscious and that animal cruelty is not okay. But it's okay to cause pain and suffering to the avatar in your computer, at least today. That may not be the case 20 years from now.

Artificial Intelligence & Reverse Engineering the Brain

SO: Do you think some software AI or AGI agent will pass the Turing test, possibly before a reverse-engineered brain is available in silicon?

RK: I'm working on a book called *How the Mind Works and How to Build One.* It's mostly about the brain, but the reason I call it the mind rather than the brain is to bring in these issues of consciousness. A brain becomes a mind as it melds with its own body, and in fact, our sort of circle of identity goes beyond our body. We certainly interact with our environment. It's not a clear distinction between who we are and what we are not.

My concept of the value of reverse engineering the human brain is not that you just simulate a brain in a sort of mechanistic way, without trying to understand what is going on. David Chalmers says he doesn't think this is a fruitful direction. And I would agree that just simulating a brain... *Mindlessly,* so to speak... That's not going to get you far enough. The purpose of reverse engineering the human brain is to understand the basic principles of intelligence.

Once you have a simulation working, you can start modifying things. Certain things may not matter. Some things may be very critical. So you learn what's important. You learn the basic principles by which the human brain handles hierarchies and variance, properties of patterns, high-level features and so on. And it appears that the neocortex has this very uniform structure. If we learn those principles, we can then engineer them and amplify them and focus on them. That's engineering.

Now, a big evolutionary innovation with *homo sapiens* is that we have a bigger forehead so that we could fit a larger cortex. But it's still quite limited. And it's running on the substrate that transmits information from one part of the brain to another at a few hundred feet per second, which is a million times slower than electronics. The intraneural connections compute at about 100 or

200 calculation per second, which is somewhere between 1,000,000 to 10,000,000 times slower than electronics. So if we can get past the substrates, we don't have to settle for a billion of these recognizers. We could have a trillion of them, or a thousand trillion. We could have many more hierarchal levels. We can design it to solve more difficult problems.

So that's really the purpose of reverse engineering the human brain. But there are other benefits. We'll get more insight into ourselves. We'll have better means for fixing the brain. Right now, we have vague psychiatric models as to what's going on in the brain of someone with bipolar disease or schizophrenia. We'll not only understand human function, we'll understand human dysfunction. We'll have better means of fixing those problems. And moreover we'll "fix" the limitation that we all have in thinking in a very small, slow, fairly-messily organized substrate. Of course, we have to be careful. What might seem like just a messy arbitrary complexity that evolution put in may in fact be part of some real principle that we don't understand at first.

I'm not saying this is simple. But on the other hand, I had this debate with John Horgan, who wrote a critical article about my views in IEEE Spectrum. Horgan says that we would need trillions of lines of code to emulate the brain and that's far beyond the complexity that we've shown we can handle in our software. The most sophisticated software programs are only tens of millions of lines of code. But that's complete nonsense. Because, first of all, there's no way the brain could be that complicated. The design of the brain is in the genome. The genome — well... it's 800 million bytes. Well, back up and take a look at that. It's 3 billion base pairs, 6 billion bits, 800 million bytes before compression — but it's replete with redundancies. Lengthy sequences like ALU are repeated hundreds of thousands of times. In *The Singularity is Near*, I show that if you apply lossless compression, you get down to about 50 million bytes. About half of that is the brain, so that's about 25 million bytes. That's about a million lines of code. That's one derivation. You could also look at the amount of complexity that appears to be necessary to perform functional simulations of different brain regions. You actually get about the same answer, about a million lines of code. So with two very different methods, you come up with a similar order of magnitude. There just isn't trillions of lines of code — of complexity — in the design of the brain. There is trillions, or even thousands of trillions of bytes of information, but that's not complexity because there's massive redundancy.

For instance, the cerebellum, which comprises half the neurons in the brain and does some of our skill formation, has one module repeated 10 billion times with some random variation with each repetition within certain constraints. And there are only a few genes that describe the wiring of the cerebellum that comprise a few tens of thousands of bytes of design information. As we learn skills like catching a fly ball — *then* it gets filled up with trillions of bytes of information. But just like we don't need trillion of bytes of design information to design a trillion-byte memory system, there are massive redundancies and repetition and a certain amount of randomness in the implementation of the brain. It's a probabilistic fractal. If you look at the Mandelbrot set, it is an exquisitely complex design.

SO: So you're saying the initial intelligence that passes the Turing test is likely to be a reverse-engineered brain, as opposed to a software architecture that's based on weighted probabilistic analysis, genetic algorithms, and so forth?

RK: I would put it differently. We have a toolkit of AI techniques now. I actually don't draw that sharp a distinction between narrow AI techniques and AGI techniques. I mean, you can list them — markup models, different forms of neural nets and genetic algorithms, logic systems, search algorithms, learning algorithms. These are techniques. Now, they go by the label AGI. We're going to add to that tool kit as we learn how the human brain does it. And then, with more and more powerful hardware, we'll be able to put together very powerful systems.

My vision is that all the different avenues — studying individual neurons, studying brain wiring, studying brain performance, simulating the brain either by doing neuron-by-neuron simulations or functional simulations — and then, all the AI work that has nothing to do with direct emulation of the brain — it's all helping. And we get from here to there through thousands of little steps like that, not through one grand leap.

Global Warming & GNR Technologies

SO: James Lovelock, the ecologist behind the Gaia hypothesis, came out a couple of years ago with a prediction that more than 6 billion people are going to perish by the end of this century, mostly because of climate change. Do you see the GNR technologies coming on line to mitigate that kind of a catastrophe?

RK: Absolutely. Those projections are based on linear thinking, as if nothing's going to happen over the next 50 or 100 years. It's ridiculous. For example, we're applying nanotechnology to solar panels. The cost per watt of solar energy is coming down dramatically. As a result, the amount of solar energy is growing exponentially. It's doubling every two years, reliably, for the last 20 years. People ask, "Is there really enough solar energy to meet all of our energy needs?" It's actually 10,000 times more than we need. And yes you lose some with cloud cover and so forth, but we only have to capture one part in 10,000. If you put efficient solar collection panels on a small percentage of the deserts in the world, you would meet 100% of our energy needs. And there's also the same kind of progress being made on energy storage to deal with the intermittency of solar. There are only eight doublings to go before solar meet 100% of our energy needs. We're awash in sunlight and these new technologies will enable us to capture that in a clean and renewable fashion. And then, geothermal — you have the potential incredible amounts of energy.

Global warming — regardless of what you think of the models and whether or not it's been human-caused — it's only been one degree Fahrenheit in the last 100 years There just isn't a dramatic global warming so far. I think there are lots of reasons we want to move away from fossil fuels, but I would not put greenhouse gasses at the top of the list

These new energy technologies are decentralized. They're relatively safe. Solar energy, unlike say nuclear power plants and other centralized facilities, are safe from disaster and sabotage and are non-polluting. So I believe that's the future of energy, and of resource utilization in general.

The Singularity, Utopia, & Happiness

R.U. Sirius: Have any critics of your ideas offered a social critique that gives you pause?

RK: I still think Bill Joy articulated the concerns best in his *Wired* cover story of some years ago. My vision is not a utopian one. For example, I'm working with the U.S. army on developing a rapid response system for biological viruses, and that's actually the approach that I advocate — that we need to put resources and attention to the downsides. But I think we do have the scientific tools to create a rapid response system in case of a biological viral attack. It took us five years to sequence HIV; we can sequence a virus now in one day. And we could, in a matter of days, create an RNA interference medication based on sequencing a new biological virus. This is something we created to contend with software viruses. And we have a technological immune system that works quite well.

And we also need ethical standards for responsible practitioners of AI, similar to the Asilomar Guidelines for biotech, or the Forsyth Institute Guidelines for nanotech, which are based on the Asilomar Guidelines. So it's a complicated issue. We can't just come up with a simple solution and then just cross it off our worry list. On the other hand, these technologies can vastly expand our creativity. They've already democratized the tools of creativity. And they are overcoming human suffering, extending our longevity and can provide not only radical life extension but radical life *expansion.*

There's a lot of talk about existential risks. I worry that painful episodes are even more likely. You know, 60 million people were killed in WWII. That was certainly exacerbated by the powerful destructive tools that we had then. I'm fairly optimistic that we will make it through. I'm less optimistic that we can avoid painful episodes. I do think decentralized communication we have actually helps reduce violence in the world. It may not seem that way because you just turn on CNN and you've got lots of violence right in your living room. But that kind of visibility actually helps us to solve problems.

RUS: You've probably heard the phrase from critics of the Singularity — they call it the "Rapture of the Nerds." And a lot of people who are into this idea do seem to envision the Singularity as a sort of magical place where pretty much anything can happen and all your dreams come true. How do you separate your view of the Singularity from a utopian view?

RK: I don't necessarily think they are utopian. I mean, the whole thing is difficult to imagine. We have a certain level of intelligence and it's difficult to imagine what it would mean and what would happen when we vastly expand that. It would be like asking cavemen and women, "Well, gee, what would you like to have?" And they'd say, "Well, we'd like a bigger rock to keep the animals out of our cave and we'd like to keep this fire from burning out?" And you'd say, "Well, don't you want a good web site? What about your Second Life habitat?" They couldn't imagine these things. And those are just technological innovations

So the future does seem magical. But that gets back to that Arthur C. Clark quote that any sufficiently developed technology is indistinguishable from magic. That's the nature of technology — it transcends limitations that exist without that technology. Television and radio seem magical — you have these waves going through the air, and they're invisible, and they go at the speed of light and they carry pictures and sounds. So think of a substrate that's a million times faster. We'll be overcoming problems at a very rapid rate, and that will seem magical. But that doesn't mean it's not rooted in science and technology.

I say it's not utopian because it also introduces new problems. Artificial intelligence is the most difficult to contend with, because whereas we can articulate technical solutions to the dangers of biotech, there's no purely technical solution to a so-called unfriendly AI. We can't just say, "We'll just put this little software code subroutine in our AIs, and that'll keep them safe." I mean, it really comes down to what the goals and intentions of that artificial intelligence are. We face daunting challenges.

RUS: I think when most people think of utopia, they probably just think about everybody being happy and feeling good.

RK: I really don't think that's the goal. I think the goal has been demonstrated by the multi-billion-year history of biological evolution and the multi-thousand-year history of technological evolution. The goal is to be creative and create entities of beauty, of insight, that solve problems. I mean, for myself as an inventor, that's what makes me happy. But it's not a state that you would seek to be in at all times, because it's fleeting. It's momentary.

To sit around being happy all the time is not the goal. In fact, that's kind of a downside. Because if we were to just stimulate our pleasure centers and sit around in a morphine high at all times — that's been recognized as a downside and it ultimately leads to a profound unhappiness. We can identify things that make us unhappy. If we have diseases that rob our faculties or cause physical or emotional pain — that makes us unhappy and prevents us from having these moments of connection with another person, or a connection with an idea, then we should solve that. But happiness is not the right goal. I think it represents the cutting edge of the evolutionary condition to seek greater horizons and to always want to transcend whatever our limitations are at the time. And so it's not our nature just to sit back and be happy.

Movies & Science Fiction

RUS: You've got two films coming out, *Transcendent Man* and *The Singularity is Near*. *(Note: The films are now being shown at film festivals)* What do you think the impact will be of having those two films out in the world?

RK: Well, *Transcendent Man* has already premiered at the Tribeca film festival and it will have an international premier at the Amsterdam documentary film festival next month. There's quite a lot of interest in that, and there are discussions with distributors. So it's expected to have a theatrical release both in this country and internationally early next year. And *Singularity is Near* will follow.

Movies are a really different venue. They cover less content than a book but they have more emotional impact. It's a big world out there. No matter how many times I speak — and even with all the press coverage of all these ideas, whether it's featuring me or others — I'm impressed by how many otherwise thoughtful people still haven't heard of these idea.

I think it's important that people not just understand the Singularity, which is some decades away, but the impact right now, and in the fairly near future, of the exponential growth of information technology. It's not an obscure part of the economy and the social scene. Every new period is going to bring new opportunities and new challenges. These are the issues that people should be focusing on. It's not just the engineers who should be worrying about the downsides of biotechnology or nanotechnology, for example. And people should also understand the opportunities. And I think there are anti-technology movements that continue to spread among the intelligentsia that are actually pretty ignorant.

RUS: Do you read science fiction novels and watch science fiction television, or science fiction movies?

RK: I have seen most of the popular science fiction movies.

RUS: Any that you find particularly interesting or enjoyable?

RK: Well, one problem with a lot of science fiction — and this is particularly true of movies — is they take one change, like the human-level cyborgs in the movie *AI,* and they put it in a world that is otherwise unchanged. So in *AI,* the coffee maker is the same and the cars are the same. There's no virtual reality, but you had human-level cyborgs. Part of the reason for that is the limitation of the form. To try to present a world in which everything is quite different would take the whole movie, and people wouldn't be able to follow it very easily. It's certainly a challenge to do that. I am in touch with some movie makers who want to try to do that.

I thought *The Matrix* was pretty good in its presentation of virtual reality.And they also had sort of AI-based people in that movie, so it did present a number of ideas. Some of the concepts were arbitrary as to how things work in the matrix, but it was pretty interesting.

Brain on a Chip: A Roundup of Projects Working on Silicon Intelligence

http://hplusmagazine.com/articles/ai/brain-chip

Surfdaddy Orca

Are we humans – with our carbon-based neural net "wetware" brains – at a point in history when we might be able to imprint the circuitry of the human brain using transistors on a silicon chip? A well-covered recent article in MIT's *Technology Review* reports that a team of European scientists may have taken the first steps in creating a silicon chip designed to function like a human brain.

What's involved in this seemingly Herculean task? The brain is a parallel processor. The colorful blue jay I see flitting from tree to tree in my garden appears as a single image. But the brain divides what it sees into four components: color, motion, shape, and depth. These are individually processed — *at the same time* — and compared to my stored memories (blue things, things with feathers, things that fly, other blue jays that I've seen).

My brain then combines all of these processes into one image that I see and comprehend. And that's just the vision aspect of a multiplexed moment of perception. At the same time, I smell the fragrant flowers in my garden, hear the neighbors talking about a party, feel my muscles relax as I sit in my lounge chair, and daydream about the beaches of Fiji while I answer my cell phone.

The MacBook Pro Intel core duo that I'm using to type this article is also doing several things at once. At the highest level, its world consists of programs with multiple computational threads running at the same time. Parallel processing makes programs run faster because there are more CPUs or cores running them.

Today's most powerful supercomputers are all massively parallel processing systems with names like Earth Simulator, Blue Gene, ASCI White, ASCI Red, ASCI Purple, and ASCI Thor's Hammer. Through Moore's Law — which states that the number of transistors on a chip double every eighteen months – single chips that function as parallel processor arrays are becoming cost effective. Examples include chips from Ambric, picoChip, and Tilera.

The brain is also massively parallel, but currently on a different scale than the most powerful supercomputers. The human cortex has about 22 billion neurons and 220 trillion synapses. A supercomputer capable of running a software simulation of the human brain doesn't yet exist. Researchers estimate that it would require at least a machine with a computational capacity of 36.8 petaflops (a petaflop is a thousand trillion floating point operations per second) and a memory capacity of 3.2 petabytes – a scale that supercomputer technology isn't expected to hit for at least three years.

Enter a team of scientists in Europe that has created a silicon chip designed to function like a human brain. With 200,000 neurons linked up by 50 million synaptic connections, the chip is still orders of magnitude from a human brain. Yet the chip can "mimic the brain's ability to learn more closely than any other machine" thus far.

"The chip has a fraction of the number of neurons or connections found in a brain, but its design allows it to be scaled up." So says Karlheinz Meier, a physicist at Heidelberg University in Germany, and the coordinator of the Fast Analog Computing with Emergent Transient States project, or FACETS.

Henry Markram, head of the Blue Brain project at the Ecole Polytechnique Fédérale de Lausanne, uses the same databases of neurological data as FACETS. Among the challenges he faces is "recreating the three-dimensional structure of the brain in a 2-D piece of silicon."

Markram admits that the simulations of biological brain functions using a silicon chip are still crude. "It's not a brain. It's more of a computer processor that has some of the accelerated parallel computing that the brain has," he says.

Markram doubts that the FACETS hardware approach will ultimately offer much insight into how the brain works. For example, unlike the Blue Brain project, researchers aren't able to perform drug testing – simulating the effects of drugs on the brain using silicon. "It's more a platform for artificial intelligence than understanding biology," he says.

Markram's Blue Brain project is the first comprehensive attempt to reverse engineer the mammalian brain. The brain processes information by sending electrical signals from neuron-to-neuron using the "wiring" of dendrites and axons. In the cortex, neurons are organized into basic functional units — cylindrical volumes – each containing about 10,000 neurons that are connected in an intricate but consistent way. These units operate much like microcircuits in a computer. This microcircuit, known as the neocortical column, is repeated millions of times across the cortex.

The first step of the project is to recreate this fundamental microcircuit, down to the level of biologically accurate individual neurons. The microcircuit can then be used in simulations such as a genetic variation in particular neurotransmitters, mimicking what happens when the molecular environment is altered using drugs.

Brains In Silicon, an interdisciplinary program at Stanford, also combines neurobiological research with electrical engineering. The program has two complementary objectives: to use the existing knowledge of brain function to design an affordable supercomputer that can then, itself, serve as a tool to investigate brain function, "feeding back and contributing to a fundamental, biological understanding of how the brain works."

Kwabena Boahen, Brains In Silicon principal investigator and an associate professor of bioengineering at Stanford, has been working on implementing neural architectures in silicon. One of the main challenges to building this system in hardware, explains Boahen, is that each neuron connects to others through 8,000 synapses. It takes about 20 transistors to implement a synapse. Clearly, building the silicon equivalent of 220 trillion synapses is not an easy problem to solve.

The quest to reverse-engineer the human brain is described in detail in Jeff Hawkins' well-known book *On Intelligence*. Hawkins believes computer scientists have focused too much on the end product of artificial intelligence. Like B.F. Skinner, who held that psychologists should study stimuli and responses and essentially ignore the cognitive processes that go on in the brain, he holds that scientists working in AI and neural networks have focused too much on inputs and outputs rather than the neurological system that connects them.

Hawkins' company, Numenta, is creating a new type of computing technology modeled on the structure and operation of the neocortex. The technology is called Hierarchical Temporal Memory, or HTM, and is applicable to a broad class of problems from machine vision to fraud detection to semantic analysis of text. HTM is based on the theory of the neocortex first described in Hawkins' book.

In *The Singularity Is Near*, Ray Kurzweil comments that, "...hardware computational capacity is necessary but not sufficient. Understanding the organization and content of these resources – the software of intelligence — is even more critical and is the objective of the brain reverse engineering undertaking." He goes on to famously say that once a computer achieves a human level of intelligence, it will necessarily soar past it.

H+ contributor Ben Goertzel (like Kurzweil) has stated that – given the problems facing humanity – we may not be able to wait on advances in hardware and the revers -engineering of the brain to achieve the AI vision of human-like intelligence (or greater). His Novamente Artificial General Intelligence (AGI) software is not dependent on a specific hardware architecture, although it will obviously benefit from massively parallel supercomputer architectures. Key cognitive mechanisms of the system include a probabilistic reasoning engine based on a variant of probabilistic logic and an evolutionary learning engine that is based on a synthesis of probabilistic modeling and evolutionary programming. It's a different approach than reverse engineering the brain, but one that may yield results more quickly.

With research and development converging on all fronts – hardware and software – it would seem to be only a matter of time until a brain with human-level complexity is available using a massively parallel architecture on silicon chip. Karlheinz Meier's FACETS group now plans to further scale up their chips, connecting a number of wafers to create a superchip with a total of a billion neurons and 1013 (10 trillion) synapses, well on the way to the 22 billion neurons and 220 trillion synapses of the human brain.

If Ray Kurzweil is right, superchip development won't stop at 22 billion neurons, even if Moore's law is no longer applicable and it becomes impossible to get additional transistors on a piece of silicon. Physicist Freeman Dyson at Princeton University has visualized spheres extracting usable stellar energy. Currently the stuff of SciFi, a "Class B stellar engine" would consist of a series of nested Dyson spheres – a Matryoshka brain like a series of Russian dolls enclosed inside each other – and composed of nanoscale computers powered by a star.

The Chinese Singularity

http://hplusmagazine.com/articles/ai/chinese-singularity

Ben Goertzel

Dr. Hugo de Garis, the father of evolvable hardware and a redoubtable AI researcher, moved to China several years ago, and is now leading the Artificial Brain Lab at Xiamen University. He is convinced a Singularity in the vein of Vinge and Kurzweil is likely to occur later this century — and that China is the most likely place for human-level Artificial General Intelligence (AGI) and the other critical technologies underlying the Singularity to arise.

As Hugo puts it: "China has a population of 1.3 billion. The US has a population of 0.3 billion. China has averaged an economic growth rate of about 10% over the past 3 decades. The US has averaged 3%. The Chinese government is strongly committed to heavy investment into high tech. From the above premises, one can virtually prove, as in a mathematical theorem, that China in a decade or so will be in a superior position to offer top salaries (in the rich Southeastern cities) to creative, brilliant Westerners to come to China to build artificial brains — much more than will be offered by the US and Europe. With the planet's most creative AI researchers in China, it is then almost certain that the planet's first artificial intellect to be built will have Chinese characteristics."

Is he right?

(Full disclosure: I spent a month at Hugo's lab in Xiamen this summer, and Hugo and I recently received word that the Chinese National Science Foundation has approved a grant to fund his lab to pursue some of our joint research on cognitive robotic, aimed at enabling the Nao humanoid robot to learn,

reason and communicate in English and Chinese.I've even debated making a move to Xiamen myself, so I can't claim great objectivity on this topic... and indeed it was with some personal fascination that I asked a variety of individuals involved with AI research and software technology in China whether the Singularity will be Chinese.)

My first destination on my quest for wisdom about the Chinese Singularity was a visit to Temple University AGI researcher Dr. Pei Wang, a long-time US resident who visits his home country of China each summer. Pei expressed a milder version of Hugo's sentiments: "I think China is among the most likely places (though not the only one) where the first truly/generally intelligent system will be created... Given the population size and education level of China, its chance is quite large... There are profound intellectual resources to make AGI happen."

Pei points out that "one of China's major advantages is the lack of strong skepticism about AGI resulting from past failures." The US and Japan have spent large sums on AI research in past decades with disappointing results, and as a consequence are particularly skeptical of AI relative to other research areas. China never had that experience, and is making its first serious foray into AI in an era blessed with more powerful computers and deeper knowledge of cognition and computer science. Pei also noted that the research community in China tends to favor incremental research over riskier attempts at paradigm-shifting progress. This seems to have held true in the AI field, so far: Chinese AI researchers have made important innovations in multiple areas such as fuzzy systems, genetic algorithms, machine translation and spatiotemporal logic, but haven't yet launched any AI revolutions.

Dr. Min Jiang, an assistant professor in Hugo's Artificial Brain Lab specializing in AI cognition and formal logic, indicated a factor counterbalancing this conservatism: "In many fields, China today is a follower. But maybe this is part of the reason China wants to spend research money on innovative projects. It can be considered a 'tuition fee' and an investment in the future. Even if some projects fail, we can learn lots of things from the experience." The funding Hugo's lab has received seems to be evidence for this perspective. And this spirit of experimentation is precisely what will be needed to create AGI and other radical Singularity-enabling technologies.

If the Chinese fund an experimental Singularity-relevant project, and it yields sufficiently impressive results to excite the "power circle," dramatic things might happen.

Min offered further insights into China: "I think the most important advantage (or disadvantage) is the [political and governmental] system. If the power circle thinks a project is crucial, we do that with all the strength of the country: for example — A-bomb, spacecraft." Another example is the First Solar initiative launched in September 2009, a 10-year project aimed at blanketing 25 square miles of Inner Mongolia with solar panels, generating 2 billion watts of power, enough to light up three million homes. When the Chinese government really wants to do something, they think big.

This combination — a willingness to experiment with new ideas, and a willingness to put massive funding behind selected initiatives — is intriguing. If the Chinese fund an experimental Singularity-relevant project, and it yields sufficiently impressive results to excite the "power circle," dramatic things might happen. This is exactly what Hugo has in mind with his "CABA" proposal, which he presented at the Oriental Technology Forum in Shanghai this October: "What I propose is that the Chinese government should create a 'CABA' (Chinese Artificial Brain Administration) over the next 5-10 years, consisting of thousands of scientists and engineers to design artificial brains for the Chinese home-robot industry and other applications. CABA would do for artificial brains what the CNSA (Chinese National Space Administration) does for space, i.e. it employs thousands of scientists and engineers to design and control rockets for China's space applications." Wildly ambitious? Perhaps. But so is covering 25 square miles of Mongolia with solar panels.

I found Western entrepreneurs operating technology firms in China to be the most skeptical voices regarding the possibility of a Chinese Singularity. I interviewed two such individuals in depth. Both are Singularity optimists, and both were concerned that their remarks be kept anonymous, to avoid potential harm to their Chinese business work. Both put the odds of a Chinese Singularity launch at less than 5%, and they gave similar reasons: they consider Chinese engineers on the whole "below average in problem solving and creative thinking," "very conservative, unwilling to considering doing anything that is not established practice." One of them also noted that "Local above-average talent insists on working for American, European, Japanese, or Korean (in that order) firms rather than Chinese firms. So, the best chance for AI breakthrough here is with a foreign research effort."

I have heard this complaint about a "lack of creativity" before, but it runs counter to my own experience at Xiamen University. There, while I've encountered some conservatism, I've also met some extremely creative and individualistic young professors and students. In my experience, researchers in China are just as creative as anywhere else — but there are subtle sociocultural issues at play, with different implications in the corporate and university contexts. Chinese culture, in its current incarnation, tends to spawn social structures that suppress rather than encourage the expression of personal creativity. It also doesn't tend to support Western-style teamwork. There's a proverb to the effect that "a lone Chinese is as powerful as a dragon; but three Chinese together can't match a bug." These are real issues, yet ones that can be worked around with care, using different methods depending on the context.

It must be understood that, regarding personal creativity as other matters, Chinese history has been powerfully cyclical. In his controversial recent book *1434*, Gavin Menzies argues that the Italian Renaissance was launched by a fleet of Chinese ships that sailed to Italy and distributed advanced knowledge including encyclopedias from which Leonardo da Vinci indirectly derived many of his celebrated illustrations of mechanical devices, flying machines, and so forth. Whether or not this thesis is true, Menzies presents compelling evidence regarding the advanced level of Chinese engineering and science during that time period, before a change of administration in Beijing ended the period of wild invention and exploration and brought a new era of conservatism to China. My point is that Chinese "cultural DNA" has plenty of innovation and creativity in it, and one must be careful to distinguish stable characteristics of Chinese culture from cyclically shifting ones. The pendulum of Chinese culture swings in a wide arc.

In the corporate software development context, one strategy for working around counterproductive cultural tendencies and bringing out Chinese creativity is the adoption of "agile" software development methods. A 2008 article in *InfoQ* summarized the experiences of five Chinese software firms who adopted the "Scrum" development methodology — a very dynamic teamwork-based approach to making software that requires constant adaptive creativity on the part of the participants. Three found the approach successful; two did not. Those who didn't find success complained that the development teams or managers understood the formalities but not the essence of the agile approach — the cultural disconnect was too great. And this is surely related to the reason why Chinese universities are so eager to bring in Western professors, like Hugo de Garis. It's not just the research ideas the Westerners bring— it's the different intuitions, experiences and habits regarding directing a research lab and a research program. In this sense Hugo's emphasis on China bringing "creative brilliant Westerners ... to China to build artificial brains" may be savvy. If China can leverage its economic growth and openness to innovative research directions to recruit a sufficient number of Western research mavericks, then powerful things may happen. Imagine a situation in which every Chinese city has a number of labs, focused on Singularity-relevant technologies, in which Western research leaders are hard at work bringing young Chinese scientists up to speed on Western ways of doing creative team R&D. In this quite plausible scenario, the prospect of a Chinese Singularity doesn't seem so farfetched.

Along with AGI, it's worth noting the differences between Western and Chinese attitudes on another radical future technology: life extension. Westerners tend to greet talk of immortality with skepticism or even moral disapproval — after all, the standard Christian story is that God wants us to die and go to heaven. But the Chinese memeplex is stocked with thousands of years of Taoist tales of immortality. Traditional Chinese methods of achieving immortality are often arduous; for instance Taoist Yoga has techniques involving lifelong celibacy and meditation focused on eventually giving birth to one's immortal self through the top of one's head. Many Chinese would be very open to immortality or life extension pills that could deliver the same benefits at lower cost and with greater reliability. So far this attitude has not translated into dramatic funding for life extension research, but the potential certainly is there — as is the economic motivation, since China will face a severely aging population around 2025-2030, similar to what Europe is facing now.

The Chinese government should create a Chinese Artificial Brain Administration, consisting of thousands of scientists and engineers to design artificial brains for the Chinese home-robot industry.

David Chambers of the Methuselah Foundation, discussing the 2006 Tomorrow's People Forum at Oxford University, compared Western and Chinese attitudes on life extension technology as follows: "Europeans don't look forward to a better future — but rather a managed version of the present. There's a distrust of revolutionary ideas.... [But] while Euros and Americans might have their various hang-ups about the ethics and implications of the new biology, China doesn't. Pei Xuetao, of the Beijing Institute of Transfusion Medicine [a leading institution in stem cell research and regenerative medicine], made it very clear [in his talk at the Tomorrow's People Forum] that China is open for business." Alongside research aimed at curing cancer and other diseases, Xuetao and his colleagues have made important discoveries involving cellular senescence and apoptosis, working toward an understanding of the genetic networks that make us age.

These differing attitudes toward immortality may be connected with attitudes toward AGI. Western skepticism about AI may not be entirely due to prior AI funding fiascos, but may also be tied to deep-seated cultural issues. The same Christian memes that tell us we're supposed to die and go to heaven also tell us that machines can never truly be conscious because they lack an immortal soul. Yet Changle Zhou, the dean who supervises de Garis's Artificial Brain Project, regularly refers to Hugo's work as the "Conscious Robot Project." Chinese culture has little of the West's subliminal resistance to thinking machines or immortal people and this cultural difference may manifest itself in the next decades in subtle ways.

Another cultural difference to remember is that extrapolating progress in China by plotting linear or exponential curves often doesn't make sense. Progress in China often matches the biological notion of "punctuated equilibrium" — long periods of relative stability punctuated by surprising and sudden changes. The Cultural Revolution and the recent shift to market-oriented "Socialism with Chinese characteristics" illustrate this phenomenon — as do the sudden initiation and cessation of Chinese global seafaring in the 1400s, and dozens of other instances in China's long history. It's easy to imagine a single technological breakthrough catalyzing one of these sudden shifts in the near future. It could be intelligent robotics, it could be life extension or something else wild and unforeseen. While this article was in the editing process, I heard some fascinating talk about a very substantial amount of funding being allocated by Beijing to a project called the "head brain instrument" (three Chinese characters) intended to improve neural function and hence accelerate human learning. I don't know enough about it to assess the viability but if it works out, it sounds like the sort of thing that could punctuate any nation's equilibrium!

The possibility of a Chinese Singularity may strike fear into the hearts of American nationalists or Eurocentrists, but it's not clear that it will make a big difference which nation makes the crucial breakthroughs. In today's scientific world "information wants to be free" — and since the most likely path to a Chinese Singularity involves collaboration of Chinese and Western researchers, the odds of an insular Chinese Singularity uniquely serving Chinese national interests seem fairly low. The work in Hugo's lab in Xiamen centers on open-source software development. It's evolving cooperatively with work done by AI coders outside China and it's delivered freely to the international research community.

So what's the verdict? Given China's lack of hang-ups about AGI and life extension, its powerful economic growth, its large population of smart and hard-working young scientists, and its eagerness to import western research leaders — will the Singularity be launched in China? I'll give the last words to two creative young scientists from Xiamen University.

Min Jiang made a statement I found intriguing given China's ongoing obsession with its 5000-year-old culture: "Today's China is a young boy, and as you know, eighteen is the age full of curiosity and fantasy about the future!" And Ruiting Lian, a PhD student at the Artificial Brain lab focused on multilingual natural language comprehension, generation and dialogue, cut to the chase more directly: "In China, the best answer to every question is *maybe*".

Build an Optimal Scientist, Then Retire

An interview with AI scientist Jürgen Schmidhuber

http://hplusmagazine.com/articles/ai/build-optimal-scientist-then-retire

Michael Anissimov

Jürgen Schmidhuber (pronounced Yirgan Shmidhoobuh) is one of the world's most interesting minds in artificial intelligence. Schmidhuber is co-director of the Swiss AI lab IDSIA in Lugano and a professor of Cognitive Robotics at the Tech University Munich. Since the 1980s, he has worked on topics in computer science and robotics including artificial curiosity, theories of surprise, incremental program evolution ("the first approach that made it possible to evolve entire soccer team strategies from scratch", according to his website), universal learning algorithms, optimally self-improving theoretical constructs called Gödel machines, artificial ants, robots that are taught how to tie shoelaces using reinforcement learning, and much more. A search for Jürgen on Google Scholar returns over 4,000 results.

At a recent talk at Singularity Summit 2009 in New York, a gathering of futurists and researchers from cutting-edge fields including AI, nanotech, and biotech, Dr. Schmidhuber touched the surface of some of his research interests, including a tongue-in-cheek argument that the Singularity must occur in 1540, based on a seemingly accelerating trend of major events that occurred between 1444 and 1517.

Dr. Schmidhuber is also an artist, creating "low-complexity art" based on principles from algorithmic information theory. In this interview, I ask Dr. Schmidhuber about his work, his philosophy towards artificial intelligence, and views on the future.

H+: Your website states that your "main scientific ambition is to build an optimal scientist, then retire." What makes you think that creating generally intelligent AI will be possible in the next few decades rather than taking centuries or never occurring?

JÜRGEN SCHMIDHUBER: In the new millennium, work at IDSIA already led to theoretically optimal universal problem solvers, such as the asymptotically fastest algorithm for all well-defined problems, and the Gödel Machine. AI is becoming a formal science! The basic principles of the new methods are very simple. This makes me optimistic that the answer to an essential remaining open question is also simple: If an intelligent agent can execute only a fixed number of computational instructions per unit time interval (say, 10 trillion elementary operations per second), what is the best way of using them to get as close as possible to the recent theoretical limits of universal AIs?

H+: The website for the Dalle Molle Institute for AI, which you co-direct, lists dozens of fascinating projects. Can you tell us a little bit about which projects you are working on currently?***

JS: We have several projects on brain-like recurrent neural nets (RNN) -- networks of neurons with feedback connections. Biological RNN can learn many behaviors/sequence processing tasks/algorithms/programs that are not learnable by traditional machine learning methods. This explains the rapidly growing interest in artificial RNN for technical applications: general computers that can learn algorithms to map input sequences to output sequences, with or without a teacher. They are computationally more powerful and biologically more plausible than other adaptive approaches such as Hidden Markov Models (no continuous internal states), feedforward networks and Support Vector Machines (no internal states at all). Our artificial RNN have recently given state-of-the-art results in time series prediction, adaptive robotics and control, connected handwriting recognition, image classification, aspects of speech recognition, protein analysis, stock market prediction, and other sequence learning problems. We are continuing to improve them (see resources).

We also have ongoing projects based on a simple principle explaining essential aspects of subjective beauty, novelty, surprise, interestingness, attention, curiosity, creativity, music, jokes, and art & science in general. Any data becomes temporarily interesting by itself to some self-improving but computationally limited subjective observer once he learns to predict or compress the data in a better way, thus making it subjectively simpler and more "beautiful." Curiosity is the desire to create or discover more non-random, non-arbitrary, regular data that is novel and surprising not in the traditional sense of Boltzmann and Shannon but in the sense that it allows for compression progress because its regularity was not yet known. This drive maximizes interestingness, the first derivative of subjective beauty or compressibility... that is, the steepness of the learning curve. It motivates exploring infants, pure mathematicians, composers, artists, dancers, comedians, yourself, and (since 1990) our increasingly complex artificial systems. Ongoing project: build artificial robotic scientists and artists equipped with curiosity and creativity (see resources).

H+: What is the "asymptotically fastest algorithm for all well-defined problems?" Maybe you could expand on that somewhat for non-technical readers?

JS: At IDSIA my former postdoc Marcus Hutter (now professor in Canberra) wrote down an algorithm that takes any formally well-defined problem (say, a Traveling Salesman Problem or whatever) as an input and solves it as quickly as the unknown fastest program that provably solves all instances of the given problem class (all TSPs in our example), save for a very small multiplicative slowdown (1% or less) and an additive constant that does not depend on the problem size (the number of cities in our example). Most problems are so big that the constant becomes totally negligible. Is that the end of computer science? Almost but not quite. Our universe is full of small problems where the additive constant is still relevant. However, the self-referential

Gödel machine (also developed at IDSIA) can deal with such constants in a way that is again theoretically optimal in a sense.

H+: You work on both biologically inspired and more formal theoretical approaches to AI. Would you call yourself a "neat" or a "scruffy" with respect to AI?

JS: I am promoting the New AI, that is, AI as a Formal Science, as opposed to a bunch of heuristics. Heuristics come and go, theorems are for eternity. The new millennium results of IDSIA describe the first general AIs that are provably theoretically optimal in various important senses. However, I am ready to admit that inspiration for the neat and mathematically rigorous systems often comes from scruffy biological systems. In fact, several of our most successful practical AI systems are not based on the recent theoretical optimality results. I believe, however, that theory and practice will converge soon.

H+: What do you think of using virtual worlds vs. real-world robotics to train AI systems? Is there a major difference?

JS: Answer A... In our research, virtual and real worlds actually complement each other. We use machine learning and artificial curiosity to learn or improve simulations of the real world, then train the robot in the sim to achieve desirable goals (mental trials can be much faster and safer than real trials). Then we transfer the learned behavior back to the real robot, and so on. Problem: current hardware is too slow when it comes to modeling very complex robots in very complex environments.

Answer B... No difference to the extent that the real world itself may be just a sim. Neither Heisenberg's uncertainty principle nor Bell's inequality exclude the possibility, that the Universe, including all observers inhabiting it, is in principle computable by a completely deterministic computer program, as first suggested by computer pioneer Konrad Zuse in 1967. Then the simplest explanation of our universe is the simplest program that computes it. In 1997, I pointed out that the simplest such program actually computes all possible universes with all types of physical constants and laws, not just ours. More papers on this can be found on the IDSIA site (see resources).

H+: You're known for designing something called a Gödel machine. Can you tell us a little bit about what that does?

JS: It's a self-referential universal problem solver. In 1931, Gödel exhibited the limits of mathematics and computation by creating a formula that speaks about itself, claiming to be unprovable by an algorithmic theorem prover: either the formula is true but unprovable, or math itself is flawed in an algorithmic sense. This inspired my Gödel machine -- an agent-controlling program that speaks about itself, ready to rewrite itself in arbitrary fashion once it has found a proof that the rewrite is useful according to an arbitrary user-defined utility function (all well-defined problems can be encoded by such a utility function). Any self-rewrite of the Gödel machine is necessarily globally optimal – no local maxima! – since this proof necessarily must have demonstrated the uselessness of continuing the proof search for even better rewrites. A Gödel machine will optimally speed up its proof searcher and other program parts, provided the speed up's utility is indeed provable. More papers on this can be found on the IDSIA Gödel Machine page.

H+: In your excellent talk at the Singularity Summit 2009, you described simple algorithmic principles that underlie discovery, subjective beauty, selective attention, curiosity and creativity. What are those principles?

JS: They are very simple indeed. All we need is (1) An adaptive predictor or compressor of the continually growing sensory data history, reflecting what's currently known about sequences of actions and sensory inputs, (2) A learning algorithm (e.g., a recurrent neural network algorithm) that continually improves the predictor or compressor (detecting novel spatio-temporal patterns that subsequently become known patterns), (3) Intrinsic rewards measuring the predictor's or compressor's improvements due to the learning algorithm, (4) A reward optimizer or reinforcement learner that translates those rewards into action sequences expected to optimize future reward, thus motivating the agent to create additional novel patterns predictable or compressible in previously unknown ways.

We implemented / discussed the following variants:

1. Intrinsic reward as measured by improvement in mean squared error (1991).
2. Intrinsic reward as measured by relative entropies between the agent's priors and posteriors (1995).
3. Learning of probabilistic, hierarchical programs and skills through zero-sum intrinsic reward games (1997-2002).
4. Mathematically optimal, intrinsically motivated systems driven by compression progress (2006-2009).

How does the theory informally explain the motivation to create or perceive art and music? For example, why are some songs interesting to some observer? Not the song he just heard ten times in a row. It became too predictable in the process. Not the other weird one with the completely unfamiliar rhythm and tonality. It seems too irregular and contains too much arbitrariness and subjective noise. The observer is interested in songs that are unfamiliar enough to contain somewhat unexpected harmonies or melodies or beats etc., but familiar enough to allow for quickly recognizing the presence of a new learnable regularity or compressibility in the sound stream: a novel pattern! Sure, this song will get boring over time, but not yet.

All of this perfectly fits our principle: the current compressor of the observer tries to compress his history of acoustic and other inputs where possible. The action selector tries to find history-influencing actions such that the continually growing historic data allows for improving the compressor's performance. The interesting musical and other subsequences are precisely those with previously unknown yet learnable types of regularities, because they lead to compressor improvements. The boring patterns are those that are either already perfectly known or arbitrary or random, or whose structure seems too hard to understand. Similar statements not only hold for other dynamic art including film and dance (taking into account the compressibility of controller action sequences), but also for painting and sculpture, which also cause dynamic pattern sequences due to attention-shifting actions of the observer.

How does the theory explain the nature of inductive sciences such as physics? If the history of the entire universe were computable, and there is no evidence against this possibility, then its simplest explanation would be the shortest program that computes it. Unfortunately there is no general way of finding the shortest program computing any given data. Therefore physicists have traditionally proceeded incrementally, analyzing just a small aspect of the world at any given time, trying to find simple laws that allow for describing their limited observations better than the best previously known law, essentially trying to find a program that compresses the observed data better than the best previously known program. An unusually large compression breakthrough deserves the name discovery. For example, Newton's law of gravity can be formulated as a short piece of code that allows for substantially compressing many observation sequences involving falling apples and other objects. Although its predictive power is limited – for example, it does not explain quantum fluctuations of apple atoms – it still allows for greatly reducing the number of bits required to encode the data stream by assigning short codes to events that are predictable with high probability under the assumption that the law holds. Einstein's general relativity theory yields additional compression progress as it compactly explains many previously unexplained deviations from Newton's predictions. Most physicists believe there is still room for further advances, and this is what is driving them to invent new experiments unveiling novel, previously unpublished patterns. Physicists are just following their compression progress drive!

We have ongoing projects based on a simple principle explaining essential aspects of subjective beauty, novelty, surprise, interestingness, attention, curiosity, creativity, music, jokes...

How does the compression progress drive explain humor? Some subjective observers who read a given joke for the first time may think it is funny. Why? As the eyes are sequentially scanning the text the brain receives a complex visual input stream. The latter is subjectively partially compressible as it relates to the observer's previous knowledge about letters and words. That is, given the reader's current knowledge and current compressor, the raw data can be encoded by fewer bits than required to store random data of the same size. The punch line at the end, however, is unexpected. Initially this failed expectation results in sub-optimal data compression — storage of expected events does not cost anything, but deviations from predictions require extra bits to encode them. The compressor, however, does not stay the same forever. Within a short time interval, its learning algorithm improves its performance on the data seen so far, by discovering the non-random, non-arbitrary and therefore compressible pattern relating the punch line to previous text and previous knowledge of the reader. This saves a few bits of storage. The number of saved bits (or a similar measure of learning progress) becomes the observer's intrinsic reward, possibly strong enough to motivate him to read on in search for more reward through additional yet unknown patterns. The recent joke, however, will never be novel or funny again.

H+: If intelligent machines were created tomorrow, what sort of implications do you think that would have for humanity and civilization?

JS: Gödel machines and the like will rapidly improve themselves and become incomprehensible. It's a bit like asking an ant of 10 million years ago: If humans were created tomorrow, what sort of implications do you think that would have for all the ant colonies? In hindsight we know that many ant colonies are still doing fine, but some of them (for example, those in my house) have goal conflicts with humans, and live dangerously.

Here Come the Neurobots:
Brain Bots are Developing Personalities –
and a Whole Lot More

http://hplusmagazine.com/articles/ai/here-come-neurobots

Steve Kotler

Can we build a brain from the ground up, one neuron (or so) at a time? That's the goal of neurobotics, a science that sits at the convergence of robotics, artificial intelligence, computer science, neuroscience, cognitive psychology, physiology, mathematics and several different engineering disciplines. Computationally demanding and requiring a long view and a macroscopic perspective (qualities not often found in our world of impatient specialization), the field is so fundamentally challenging that there are only around five labs pursuing it worldwide.

Neurobotics is an outgrowth of a growing realization that, when it comes to understanding the brain, neither computer simulations nor top-down robotic models are getting anywhere close. As Dartmouth neuroscientist and Director of the Brain Engineering Lab Richard Granger puts it, "The history of top-down-only approaches is spectacular failure. We learned a ton, but mainly we learned these approaches don't work."

Gerald Edelman, a Nobel Prize-winning neuroscientist and Chairman of Neurobiology at Scripps Research Institute, first described the neurobotics

approach back in 1978. In his "Theory of Neuronal Group Selection," Edelman essentially argued that any individual's nervous system employs a selection system similar to natural selection, though operating with a different mechanism. "It's obvious that the brain is a huge population of individual neurons," says UC Irvine neuroscientist Jeff Krichmar. "Neuronal Group Selection meant we could apply population models to neuroscience, we could examine things at a systems' level." This systems approach became the architectural blueprint for moving neurobotics forward.

The Edge of Real Brain Complexity

The robots in Jeff Krichmar's lab don't look like much. CARL-1, his latest model, is a squat, white trash can contraption with a couple of shopping cart wheels bolted to its side, a video camera wired to the lid, and a couple of bunny ears taped on for good measure. But open up that lid and you'll find something remarkable — the beginnings of a truly biological nervous system. CARL-1 has thousands of neurons and millions of synapses that, he says, "are just about the edge of the amount of size and complexity found in real brains." Not surprisingly, robots built this way — using the same operating principles as our nervous system — are called *neurobots*.

Krichmar emphasizes that these artificial nervous systems are based upon neurobiological principles rather than computer models of how intelligence works. The first of those principles, as he describes it, is: "The brain is embodied in the body and the body is embedded in the environment — so we build brains and then we put these brains in bodies and then we let these bodies loose in an environment to see what happens." This has become something of a foundational principle — and the great and complex challenge — of neurobotics.

When you embed a brain in a body, you get behavior not often found in other robots. Brain bots don't work like Aibo. You can buy a thousand different Aibos and they all behave the same. But brain bots, like real brains, learn through trial and error, and that changes things. "Put a couple of my robots inside a maze," says Krichmar, "let them run it a few times, and what each of those robots learns will be different. Those differences are magnified into behavior pretty quickly." When psychologists define personality, it's along the lines of "idiosyncratic behavior that's predictive of future behavior." What Krichmar is saying is that his brain bots are developing personalities — and they're doing it pretty quickly.

Krichmar's bots develop personalities because, instead of preprogramming behaviors, these robots have neuro-modulatory systems or value judgment systems — move towards something good, move away from something bad — that are modeled around the human's dopaminergic system (for wanting or reward-based behaviors) and the noradrenergic system (for vigilance and surprise). When something salient occurs — in CARL-1's case that's usually bumping into a sensor in a maze — a signal is sent to its brain telling the bot to react to the event and remember the context for later. This is conditional learning and it mimics what occurs in real brains. It also allows Krichmar to examine one of the great puzzles in systems neuroscience — how do the brain's neurons work together?

"We're pretty sure you need a certain brain size for the level of complexity we see in biological organisms," he says, "but we don't have the tools to make a network that big behave in any stable way. The biological brain is remarkably stable. We can alter it with drugs, we can put it into all sorts of varied environments... Pretty much it still knows how to function. Our robots are still brittle by comparison."

CARL-1 has thousands of neurons and millions of synapses that "are just about the edge of the amount of size and complexity found in real brains."

Besides personality, another thing these robots develop are types of episodic and categorical memory not found in other computers. After running early brain bots Darwin X and Darwin XI through a few mazes, Edelman, working alongside Krichmar and a researcher named Jason Fleischer, found they'd naturally developed place cells — meaning they didn't program them in. These are cells in the Hippocampus that fire whenever an animal passes through a specific location, essentially linking place with time. More than that, when Edelman examined his bots' brains, he found these place cells would not only fire based on where the robot had been, but also on where it was planning to go, "which," says Krichmar, "is exactly what you would see in the brain of a rat and nothing anyone's seen in a robot before."

The Biggest Dragon: Higher Cortical Functions

Meanwhile, Richard Granger is using brain bots to hunt down yet another grail: where language originates in the brain. "It's been pretty widely demonstrated that the brain is modular and highly uniform," he says. "There are certain broad stroke differences between humans and other animals, but we can count the number of those on two hands. Yet humans can speak and animals can't. That's a pretty big difference. And even the variations that have been found in brain language areas like Broca's Area don't hint at how language could emerge from the changes found. So where is language? We've spent billions trying to track down its origins and still can't find it."

Granger believes that the only real differences between animal and human brains are size and connectivity, an argument he lays out in his book *Big Brain*. "Humans have a lot bigger brains so we have much more space for neurons to make connections, to link with other neurons. " It's in that space, in those extra connections, where Granger thinks language emerges. If he's right, as his bot brains draw closer and closer is size and complexity to human brains, language should start to emerge — and Granger will get to watch it happen.

Of course, since neurobotics is a dragon-slayer's approach, there are also a few scientists going after the biggest dragon. Just like Granger is upping complexity to examine language, researchers at Imperial College in London are doing the same thing for consciousness. "All of this work is comparable," says Granger, "because we're all modeling cortical structures to build whole brain models with the intention of seeing if higher functions like language and consciousness develop." And if what they've discovered so far is any indication, then when it comes to developing higher cortical function in neurobots, it's really not a question of if, only "when."

Can "Terminators" Actually Be Our Salvation

A Conversation with Peter Asaro

http://hplusmagazine.com/2009/05/19/can-terminators-actually-be-our-salvation/

R.U. Sirius & Surfdaddy Orca

In a fascinating paper entitled "How Just Could a Robot War Be?" philosopher Peter Asaro of Rutgers University explores a number of robot war scenarios.

Asaro imagines a situation in which a nation is taken over by robots — a sort of revolution or civil war. Would a third party nation have a just cause for interceding to prevent this?

Asaro concludes that the use of autonomous technologies such as robot soldiers is neither "completely morally acceptable nor completely morally unacceptable" according to the just war theory formulated by Michael Walzer.

Just war theory defines the principles underlying most of the international laws regulating warfare, including the Geneva and Hague Conventions. Walzer's classic book *Just and Unjust Wars* was a standard text at the West Point Military Academy for many years, although it was recently removed from the required reading list.

Asaro asserts that robotic technology, like all military force, could be just or unjust, depending on the situation.

H+: We're using semi-autonomous robots now in Iraq and, of course, we've been using smart bombs for some time now. What is the tipping point? At what point does a war become a "robot war"?

Peter Asaro: There are many kinds of technologies being used already by the U.S. military, and I think it is quite easy to see the U.S. military as being a technological system. I wouldn't call it robotic yet, though, as I think there is something important about having a "human-in-the-loop," even if the military is trying to train soldiers to behave "robotically" and follow orders without question.

I think there is always a chance that a soldier will question a bad order, even if they are trained not to, and there is a lot of pressure on them to obey.

Ron Arkin is a roboticist at Georgia Tech who has designed an architecture for lethal robots that allows them to question their orders. He thinks we can actually make robots super-moral, and thereby reduce civilian casualties and war crimes.

We might be able to design robotic soldiers that could be more ethical than human soldiers.

I think Ron has made a good start on the kinds of technological design that might make this possible. The real technical and practical challenges are in properly identifying soldiers and civilians.

The criteria for doing this are obscure, and humans often make mistakes because information is ambiguous, incomplete, and uncertain. A robot and its computer might be able to do what is optimal in such a situation, but that might not be much better than what humans can do.

More importantly, human soldiers have the capacity to understand complex social situations, even if they often make mistakes because of a lack of cultural understanding.

I think we are a long way from achieving this with a computer, which at best will be using simplified models and making numerous potentially hazardous assumptions about the people they are deciding whether or not to kill.

Also, while it would surely be better if no soldiers were killed, having the technological ability to fight a war without casualties would certainly make it easier to wage unjust and imperial wars. This is not the only constraint, but it is probably the strongest one in domestic U.S. politics of the past 40 years or so.

By the way, I see robots primarily as a way to reduce the number of soldiers needed to fight a war. I don't see them improving the capabilities of the military, but rather just automating them. The military hold an ideal vision of itself as operating like a well-oiled machine, so it seems that it can be rationalized and automated and roboticized. The reality is that the [human] military is a complex socio-technical system, and the social structure does a lot of hidden work in regulating the system and making it work well. Eliminating it altogether holds a lot of hidden dangers.

H+: Does robotic warfare heighten the possibility of accidental war, or might it guard against it?

PA: There was a news item March 2008 about a unit of the Swiss Army, about 170 infantry soldiers, entering into Liechtenstein at night by way of a dark forest. This turned out to be an accident –- they were lost during a training exercise –- so there wound up being no international incident. If there had been tensions between the countries, there could have been a just cause for Liechtenstein to declare war on Switzerland on the basis of an aggression.

Of course, Liechtenstein does not even have an army. But something similar happened in 2002 when a platoon of British Royal marines accidently invaded a Spanish beach, instead of Gibraltar.

I think the same is true of machines. They could inadvertently start a war, though this depends both on the technology malfunctioning and on the human political leadership desiring a war. Many wars have been started on false pretenses, or misconstrued or inadvertent acts: consider the sinking of the Maine in Havana or the Gulf of Tonkin incident.

H+: You talk about the notion that robots could have moral agency — even superior moral agency –- to human soldiers. What military would build such a soldier? Wouldn't such a solider be likely to start overruling the military commanders on policy decisions?

PA: I think there are varying degrees of moral agency, ranging from amoral agents to fully autonomous moral agents. Our current robots are between these extremes, though they definitely have the potential to improve.

I think we are now starting to see robots that are capable of taking morally significant actions, and we're beginning to see the design of systems that choose these actions based on moral reasoning. In this sense, they are moral, but not really autonomous because they are not coming up with the morality themselves... or for themselves.

They are a long way from being Kantian moral agents –- like some humans –- who are asserting and engaging their moral autonomy through their moral deliberations and choices. [*ed: Philosopher Immanuel Kant's "categorical imperative" is the standard of rationality from which moral requirements are derived.*] We might be able to design robotic soldiers that could be more ethical than human soldiers.

Robots might be better at distinguishing civilians from combatants; or at choosing targets with lower risk of collateral damage, or understanding the implications of their actions. Or they might even be programmed with cultural or linguistic knowledge that is impractical to train every human soldier to understand.

Ron Arkin thinks we can design machines like this. He also thinks that because robots can be programmed to be more inclined to self-sacrifice, they will also be able to avoid making overly hasty decisions without enough information. Ron also designed architecture for robots to override their orders

when they see them as being in conflict with humanitarian laws or the rules of engagement. I think this is possible in principle, but only if we really invest time and effort into ensuring that robots really do act this way. So the question is how to get the military to do this.

It does seem like a hard sell to convince the military to build robots that might disobey orders. But they actually do tell soldiers to disobey illegal orders. The problem is that there are usually strong social and psychological pressures on soldiers to obey their commanders, so they usually carry them out anyway. The laws of war generally only hold commanders responsible for war crimes for this reason. For a killing in war to truly be just, then the one doing the killing must actually be on the just side in the war. In other words, the combatants do not have equal liability to be killed in war. For a robot to be really sure that any act of killing is just, it would first have to be sure that it was fighting for a just cause. It would have to question the nature of the war it is fighting in and it would need to understand international politics and so forth.

The robots would need to be more knowledgeable than most of the high school graduates who currently get recruited into the military. As long as the war is just and the orders are legal, then the robot would obey, otherwise it wouldn't. I don't think we are likely to see this capability in robots any time soon.

I do think that human soldiers are very concerned about morality and ethics, as they bear most of the moral burdens of war. They are worried about the public reaction as well, and want to be sure that there are systems in place to prevent tragic events that will outrage the public. It's not impossible to try to control robot soldiers in this way. What we need is both the political will, and the technological design innovation to come together and shape a new set of international arms control agreements that ensures that all lethal robots will be required to have these types of ethical control systems.

Of course, there are also issues of proliferation, verification and enforcement for any such arms control strategy. There is also the problem of generating the political will for these controls. I think that robotic armies probably have the potential to change the geo-political balance of power in ways far more dramatic than nuclear arms.

We will have to come up with some very innovative strategies to contain and control them. I believe that it is very important that we are not naive about what the implications of developing robotic soldiers will mean for civil society.

Chronic Citizen:
Jonathan Lethem on P.K. Dick

Why Novels are a Weird Technology,
and Constructed Realities

http://hplusmagazine.com/articles/art-
entertainment/chronic-citizen-jonathan-lethem-
pk-dick-why-novels-are-weird-technology-a

Erik Davis

While mainstream literary figures sometimes praise their fellow writers, rarely do they present themselves publicly as hardcore pop culture fans. Since the publication of his novels *Motherless Brooklyn* and *Fortress of Solitude,* as well as his reception of the MacArthur Fellowship in 2005, Jonathan Lethem has become a successful and widely-praised author of playful and intelligent literary fictions. He has also become probably the most visible fan and proponent of the science fiction of Philip K. Dick. A few years ago, Lethem was commissioned by the august Library of America to edit a volume of Dick's writings for the publisher's definitive canon of American letters. The initial volume, *Philip K. Dick: Four Novels of the 1960s* was the best-selling title out of the gate in the history of the library, and two more Lethem-edited volumes of Dick's work followed (*Philip K. Dick: Five Novels of the 1960s & 70s* and *Philip K. Dick: VALIS and Later Novels*).

Lethem began his own writing career drawing heavily from genre fiction, both SF and hard-boiled detective novels. But he avoided getting stuck in what some SF writers refer to as "the golden ghetto," and his later work achieved mainstream recognition for more realistic, psychological, and crisply detailed tales largely rooted in a slightly altered version of the Tri-State area that is his home. His latest book, *Chronic City*, is a dark and druggy take on Manhattan — an anxious, funny, and disturbingly charming book infused with cannabis, conspiracies, astronauts, nihilistic artists, virtual objects, and pop culture mania. Though very much written in Lethem's mature voice, the book is also infused with the spirit of Philip K. Dick, who remains Lethem's first and most important influence.'

Erik Davis: How did you first encounter Philip Dick?

JONATHAN LETHEM: I first saw his books in my friend Carl's house. His dad was a science fiction fan. I was already reading the old classics — the Heinlein and Asimov and Bradbury that were on my mother's shelves. But those books were written and packaged in a style that was very '40s and '50s. And these Philip K. Dick paperbacks from the '70s looked like a whole other flavor of stuff.

The first ones I saw were *A Scanner Darkly* and the Bantam reissues of *Ubik* and *A Maze of Death*. And I just immediately connected it with psychedelia and was drawn to it. I was thirteen or fourteen when I was devouring his work and just wanting to read as much of it as I could. By the time I was about eighteen, I had read every Dick book that had been published to that point. One way or another, I found them. I just identified with him totally, and it rearranged my thinking.

I moved to the Bay Area. It was like the husk of a plan, to go and meet him. But he died, so I went anyway. I tried to look for meaningful traces, including hanging out with Paul Williams and helping him with the Philip K. Dick Society. So it was a very shaping obsessing. I kind of apprenticed myself to the guild of Philip K. Dick.

ED: As a pop culture fan, and an intense Phil Dick fan, I find it incredibly satisfying that this author that I've loved since I was a teenager is now getting his props. On the other hand, I cannot deny that there is something a bit sad about losing the esotericism of the cult. You yourself are a true fan, but one who has been instrumental in Dick's current canonization. How do you feel about that?

JL: I have a very divided conscience. I mean, just as an eyewitness, it's something to be incredibly proud of. It's almost unprecedented: the creation of a real canonical literary reputation when the person is dead and out of print, and when there was a pretty definite ceiling on how far he'd ever gotten while he was still alive and in print.

We turn them [our media] into advertisements for ourselves, rather than opportunities for shedding ourselves.

For people familiar with Dick's personal experiences, his biography and his temperament, the ironies in that are deep and bitter and complicated. You inevitably think: if he'd been alive, he would've screwed this up. He would've found some way to make it impossible that he could be treated with such simple reverence, because he was so distrustful of any form of institutional authority. He had a particularly deep, bitter and twisted suspiciousness about traditional literary authority and about academia. And frankly, to some extent, it's academia that's driven his acceptance in a canon.

When I was a kid and I discovered Philip K. Dick, I felt that I'd made this kind of soul mate contact with his work. It's a defining experience, and it feels like it's innate. For me, that experience was absolutely bound up in finding these books that were out of print. The books almost seemed like fictional artifacts. I couldn't believe there was such a writer. I still remember thinking his name seemed weird or that his titles seemed preposterous to me. It was like a secret reality unfolding in my life.

There's something about the essence of his writing that creates that feeling. And I think it's still creating it for let's say the 14-year-old equivalent of Erik Davis or Jonathan Lethem, who's discovering this book in the shiny expensive Vintage paperback editions. I still think there's something innately self-marginalizing, self-cultifying (if that's a word) about the writing. You feel like you're the only one who understands it, and he's the only one who understands you. It's like a cognitive version of a love affair.

You're making this intimate connection with this other mind. He projected that into the work.

ED: In a way, Dick is the ideal highbrow-lowbrow saint. The academics will analyze the social critique, the metafiction, the dense weave of allusions, the importance of the themes as they relate to emerging problems of simulation and consciousness and existential anomie. And at the same time, there's a pop level that's most obviously manifested in cinema, a steady stream of Hollywood films which are mostly pretty corny. And both those levels of recognition have shaped the context that allowed the Library of America to say, "Okay. This guy gets the canon badge."

But because I'm one of these cultists, I have to believe that there's something more to it. It's because his books say something about our time, even more, in some ways, than his time — the '60s and '70s. Why are we hearing these books now? What are these books telling us?

JL: I've always agreed with the view that — with science fiction — its predictive powers were the least important or least relevant aspect of its public profile. I always loved stuff like Orwell's *1984*, where he explicitly said "It's 1948, reversed." I liked writers that were doing allegorical, satirical, fantastical versions of everyday life.

That suggests that Dick's work is dated to the '60s and '70s. And I thought of him very much in this framework, and not as an extrapolative writer. He certainly doesn't have that kind of rigor or scientific chops that you find with someone like Bruce Sterling. But I think that Dick saw the makings of the contemporary reality we experience so profoundly. And this speaks to the different layers of reality in his work — the way time moves at one clip according to the calendar, but other ways in terms of mental time, psychological time, social time, American historical time. Like if you look at the terms of this absurd, hysterical healthcare debate — it's basically McCarthyism again, the Red Scare. "Socialism is coming to get us."

Dick looked around his world with a kind of skinlessness. He existed in the world and it just permeated him. Mid-'50s America was overwhelmingly alive in his vision, in such a way that he saw it simultaneously as a present and as a future. He saw the makings of the late capitalist experience embedded in that mid-century triumphalist post-war moment. And it's as though he experienced it all, in all its absurdity and its tragedy, as this overwhelming vision. And he just jotted it down as frantically as he could. And the books are so raw with that perception that they still feel like a desperate attempt to record an arriving moment. I think that's the experience of reading Philip K. Dick. He seems to be frantically trying to transcribe an arriving reality that is urgent and totally fresh.

What's missing from both the academic and pop movie descriptions you mention is that Dick is an immensely personal writer. In his own way, he's a Beat or a proto-Beat. He's like Henry Miller. One of these gargantuan, slightly egotistical but insecure, garrulous personas that just pour themselves onto the page, and says, "Love me or hate me. This is what I feel. And these are the kind of women I find sexy. And oh my god, I hate them. They're consuming me. And I feel really stupid today, but I'm going to tell you about...." And he just gives himself. And as anyone who's ever tried to write literary novels or stories or a memoir can tell you — it's not a small thing to pour yourself onto the page. And when it's accomplished, totally, you end up with the kind of monumental writers that many people find also unpleasant or toxic or unreadable.

ED: Dick set many of his tales in what we might now call a "posthuman" future of cognitive-enhancing drugs, psi powers, and other amplifications of human capacity. Many of the developments he envisioned in his own unique way are now edging ever closer to reality, and there are many enthusiasts. While not exactly bleak, Dick had a generally more dark and satiric — and often funny — take on cognitive enhancement. Was he a pessimist or a realist?

How would you characterize his particularly lesson about human for today's enthusiastic transhumans?

JL: While I'm hardly an expert on the reality of cognitive enhancement or the transhuman impulse as it's working itself out on the contemporary frontier, I suspect Dick had little to offer in the way of a "lesson" for aspirants, except in the senses that were relevant while he was coming of age as a writer — that's to say, when the breaking news in the framing of such matters involved names like Freud, Kinsey, Norbert Weiner and, well, A.E. Van Vogt. Dick's concerns were ultimately both epistemological and deeply moral — in the sense that a philosopher would use the word moral, not in the sense that, say, Joseph L. Breen would. You know, love, empathy, "what is human?" and so forth. For contemporary voyagers, these matters remain as Dick delineated them: exquisitely local, negotiated on the human-to-human, or human-to-self playing field according to an infinite number of variations and contexts. No sweeping paradigms will do here. We're all walking down the street conducting our self-Turing exams every time we pass a homeless person, or greet our spouse at the breakfast table.

ED: For proponents of the Singularity, we are on the verge of massive technological transformations that involve some version of artificial or machine intelligence. Dick had a very particular take on intelligent machines, like Joe Chip's conapt or suitcase psychiatrists. While these devices are clearly fantastic and absurd, they also express some real insight and concerns about the cultural consequences of machine intelligence. Does Dick's take seem relevant now, thirty years later? What would he say to our contemporary gadget fetishism and addiction to information machines?

JL: My best guess about such matters is that each technological transformation, up to and perhaps including the Singularity, is going to work itself out vis-à-vis "the human" according to the deep principles of all media. Defined in its largest sense, as including things like cinema, theory, drugs, computing, moving type, music, etcetera, media is utterly consciousness-transforming in ways we can no longer competently examine, given how deeply they've pervaded and altered the collective and individual consciousness that would be the only possible method for making that judgment. And yet — we still feel so utterly human to ourselves, and the proof is in the anthropomorphic homeliness that pervades the ostensibly exalted "media" in return. We humanize them, shame them, colonize and debunk them with our persistent modes of sex and neurosis and community and commerce. We turn them into advertisements for ourselves, rather than opportunities for shedding ourselves. At least so far.

ED: You're pointing toward the psychological dimension of Dick's writing. Even when you are looking at the futuristic aspect, what's really being extrapolated is a certain kind of dreamlike, subjective response to changing technological conditions. And all that is intensified by Dick's own psychological sensitivity

JL: Dick was supremely labile. He has the power to put himself, as a writer, at the mercy of his own inventions. He could construct realities and then immerse himself in them as though helpless. So he conveys the experience of the mind-altering or the reality-transforming better than nearly any writer who ever lived. As a creator of fantastical, preposterous kinds of realities that are nevertheless grounded in a critique or an insight, he was the best at two things: at making these things a kind of a reality; and then, at experiencing that reality as though it were a given. His characters — his proxies within the space of his own fictional world — are totally subsumed in it. There's no mastery exhibited. They're reading it. They're experiencing it. They're surviving it. They're not objective tour guides. His character is a sufferer who moves through these worlds.

ED: Given Dick's obsessions, it seems inevitable that he would wind up asking religious questions. These came to the head with the so-called "VALIS trilogy" he wrote toward the end of his life: *VALIS, The Divine Invasion*, **and** *The Transmigration of Timothy Archer.* **In the '70s, when he was**

read in kind of a proto-Marxist way by some critics, his later works were often dismissed at the works of a crazy man, even though the religious elements and visionary questions in his writing are evident from the get-go.

Today, some people continue to dislike these more explicitly spiritual works and prefer the more socially and critically dynamic ones of the '60s. Others see them as a crowning gesture. Did you have a sense of satisfaction in getting these three books included in the Library of America series?

JL: One of my goals was to get what I felt was the majority of Dick's masterpieces into the Library's three volumes. And for me, *VALIS* is probably one of his five greatest works. Leaving aside context, the voice, the form, the velocity, the humor, the emotion — it's a great novel. It would be a great novel in any writer's career. It had to be canonized if he was going to go into the canon. So the minute I knew we could do more books, I started scheming about how to make *VALIS* a part of the project. I like all three of the books that have been described as a trilogy, although I'm skeptical about the whole trilogy idea. Besides *VALIS*, I think *Transmigration of Timothy Archer* may be among his greatest works. And *Divine Invasion* is rock solid.

ED: With your own new book, *Chronic City*, I can very much sense the way that Dick has marked you as a reader, as a writer, as a person in the world.

JL: In the process of editing these Dick books, I felt myself recapturing a feeling of intimate kinship that came from the very beginning of wanting to be a novelist — a feeling that I wanted to, in some way, project a relationship to Dick's writing. I wanted to find a way to extend my own feelings about it into fictional space. For me, this is a book that's suffused in his influence.

ED: *Chronic City* is a dark book. What does it mean to embark on a book that, while it's entertaining and there's plenty of nice people in it that you kind of want to spend

time with, is also suffused with meditations on dread and the conundrum(s) of contemporary reality?

JL: Well, at the outset, if I started with that as a goal, I'd never do it at all. You have to start in a kind of innocence. You have to think, "I've got this funny idea." You know, "What if there was this character who didn't know he was doing such and such. And that would be fun." You start in a kind of willful naïveté about the breadth of your ambition as a survival trait — it's the only way to get in.

But I felt that this was a book, like *Fortress of Solitude*, where I wanted to disburden myself of a lot of anger. I think it's a response to living in a pretty dreadful moment — a series of dreadful moments in the last ten years. And it's a book about complicity, too — about going along with how wrong it all is because you find it entertaining or good enough or necessary, in various degrees.

ED: There is also an extraordinary amount of pot smoking in this book. Why so much?

JL: Confession compulsion? I don't know. One of the main subjects in my work is friendship, the experience of hanging out with people, of what it's like to really adore someone, argue with them, be obsessed with them — you know, compare your life to theirs, day in and day out. *Chronic City* is very much a book about friendship, and so I was trying to capture a certain vein of deep and silly and exhausted and slightly outlaw time-spending that is typified, for me, by getting high, with a certain personal group of people, again and again and again. Which isn't so much the stuff of my days right now — it can't be, you know — it's an older feeling. But it's one I hadn't ever gotten down the way I wanted to.

ED: Part of the experience I have of novels these days is that it seems like the more awake and aware and acute they are, the more they are aware of their own fragility in the face of other kinds of narrative technologies. The most obvious

example is simulation — immersive worlds that we can go into and reproduce behaviors that are more or less storylike. The fundamental character of a massive, open-ended, multi-player role-playing game is utterly different at this point than the character in a novel. How will novels stand up?

We're all walking down the street conducting our self-Turning exams everytime we pass a homeless person, or greet our spouse at the breakfast table.

JL: I'm far too close to one pole to illuminate. But I'll say that — in the face of certain kinds of rival technologies and rival frameworks for experiencing what we might call self-admitting false realities — novels are a class of virtual reality experience that has some very particular and innate bottom lines. And I happen to like those. As I see the rivals emerge, I feel that novel-making and reading becomes one option on a very large menu, and in some ways a rather antique or humble or lumpen example. But I also think some of the things that make it that are also deep strengths that are becoming more and more highlighted.

We talked about what makes Dick so compelling and personal — what made us each take him so personally when we discovered his work. And in some ways, those are elements that are innate to this very strange technology — this gigantic pile of sentences stuck between two hard covers, that someone makes this incredible commitment to read. It's a bizarre commitment, very unusual the first few times you make it — to just sit and follow, in order, each of these sentences and make the artificial reality come to life yourself by reading. It's a crazy technology, very specific and weird. Now may not be the time to take it for granted. Instead, maybe we should point out that by doing this, you do achieve a kind of weird mind meld.

ED: There are a number of Phil Dick-ian moments in *Chronic City* where we're on the edge of realizing that something we've been taking for reality is a construct or is a

convenient fiction. There's a palpable sense that recognizing this construct to its fullest extent would thrust one into the cold vacuum of space. At the same time we are immersed in more and more media constructs every day. So as we edge closer to the anxious recognition of the reality construct, there are also more technologies of distraction that try to cover that over or displace it.

JL: The reason I tend to write from the complicit point of view is I'm always struck by the deeply personal nature of the alliance we make with these opportunistic distraction mechanisms, the substitute realities that are offered to us, the way that we build ourselves into them. And that's why I always think that Dick was such an insightful writer — because he always took it personally. He was always aware of his own wish- fulfillment impulses, his own yearning to be consumed and seduced. And it's why his role as a fiction maker and as a liar was allied to his fascination and distrust of fictional realities, of marketing realities, of commercial realities and political realities — because he saw that they're rooted innately in storytelling and in emotional necessity. And that there are all sorts of things that turn out to be ideological all the way down to their bones — the family structures that we come up inside are themselves a form of storytelling, a form of myth-making and persuasion. We sell ourselves on versions of existence that are tolerable. We're all marketing.

ED: Towards the end of your book, I sense a deep ambivalence about the necessity of consoling fictions. Right next to the rage and the desire to expose the machine is a complicit adoption of conventional realities and more constructive views.

JL: Absolutely. What are the tolerances for the exposure of sustaining fictions in any given life? At some point, you're going to settle. You're going to make a snow globe and live inside it.

Isn't It Time for Cinematic Sci-Fi Television?

http://hplusmagazine.com/articles/art-
entertainment/isn%E2%80%99t-it-time-cinematic-
sci-fi-television

Kyle Munkittrick

A few weeks ago, Mark Bernardin at io9.com asked: "Why doesn't Syfy or AMC or HBO roll the dice and make a *Starship Troopers* miniseries, one with all the production value and attention to detail of *The Pacific* or *Battlestar Galactica*?" Replace Syfy with Showtime and substitute "*Starship Troopers* miniseries" with "a cinematic science-fiction show" and you have yourself a real question. While I love *Battlestar Galactica* and am leaning toward loving *Caprica*, Syfy is not exactly reliable in terms of content (e.g. WWE) or quality (e.g. *Sharktopus*). Why haven't HBO, Showtime, or AMC made a serious SF show? Maybe they just aren't getting enough good pitches. Allow me to propose a fix for the situation.

HBO, Showtime, and AMC represent the best of cinematic TV. All three have been making bank and winning oodles of awards with a very simple formula that is extremely difficult to pull off: take a genre (Gangsters, Western, Horror), complicate the cliches, focus on the rule breakers, and tell a story that is unresolved at the end of every episode. In addition to simply having a bigger budget and more liberty with violence, sex, and language, shows on HBO, Showtime, and AMC are almost always better written, filmed, and acted than anything else on television. SF as a genre is no more limiting than period dramas (*The Tudors, Mad Men, Rome*), which also require extensive costumes, elaborate sets, and an extra level of commitment from the actors. All SF does is move the timeline forward instead of backward. There is, however, one binding flaw of great SF currently on television. All the best SF shows are set in space. A savvy exec might say, "We don't need another *Star Trek* or *Battlestar* or *Stargate* — where is the human drama? Furthermore, where is the chance to talk about *our* society, *our* cities, *our* life? Space is too alien." Not to mention how Goody Two-Shoes most of the protagonists are. I mean, *Star Trek* is a utopia. Where is the *grit*?

For SF to earn its place among shows like *The Sopranos* and *Mad Men*, it needs to get dark. Thus, I present three options. The networks may battle over them as they see fit.

Core Idea: Rogue agent in cyberpunk world trying to uncover multilayer conspiracy.

Think of it as: *The Prisoner* plus *The Bourne Trilogy* plus *Blade Runner* starring a hard-boiled cyborg heroine.

Big Question: How many broken laws, hearts, and bones does it take to get to the truth?

Video games make terrible movies, but they might make spectacular television. This logic holds doubly true when the game being adapted is *Deus Ex*, perhaps one of (if not *the*) best video games of all time. A huge part of that reputation comes from the game's story. Half cyberpunk opus, half paranoid thriller, the plot of *Deus Ex* is like the *X-Files* meets *Gravity's Rainbow* meets *Rule By Secrecy* meets *Neuromancer*. By the end of the story, you're still not sure who, if anyone, was "good" and you're almost more in the dark than when you started. If there was ever a plot that could hope to rival the WTF OMG moments of *Lost*, it's definitely *Deus Ex*. Compound the absurdly complex plot with nearly every group of conspirators on the books (FEMA, MJ12, Templars, Illuminati, Freemasons) and a cyberpunk society where A.I.s roam the internet, illicit nanotech and biomods are sold behind bars and death by a police mech's laser is as likely as your next meal, and you have one hell of a back drop for whatever story you might want to tell.

Deus Ex would follow Alex Denton, a black ops UN security force agent. Denton is an army unto herself — nano-augmented and given carte blanche to maintain order by the UN, she is brutal, efficient, and all but unstoppable. The year is approximately 2045 and the UN has become a functioning world government, attempting to keep order as society decays thanks to sporadic outbreaks of the Gray Plague (imagine HIV mixed with ebola and leprosy). Elite and stoic, Denton's only real enjoyment of life come from putting down juntas that aren't UN puppets, catching hackers with their pants down, and the above-the-law privileges her job provides. When dealing with a potential bio-bomb terrorist, Denton unwittingly discovers that not only is the Gray Plague a manufactured and controlled weapon, her fellow UN agent Gunther Hermann has been the trigger man for every outbreak. Disturbed but ultimately taking a "not-my-problem" view of the situation, Denton returns to the UNSF offices but is met with bullets and EMP attacks. Mid-battle, a voice crackles over her data-link. Referring to itself as Daedalus, it helps her escape by disabling security systems, protecting her wired mind from hacks and directing her to a safe house. Her privileges, luxuries, and power gone, Denton becomes a rogue agent in a lawless world. But she has old connections, and one man, Tracer Tong, might have the information she needs to get revenge. Denton's personal vendetta and the shadow governments trying to stamp her out of existence force her to make her way through the labyrinth of the wired underworld looking for Tracer Tong and answers to questions she doesn't know.

Transmetropolitan

Core Idea: The City, a late 21st Century city filled with technology, filth, and corruption, through the eyes of Spider Jerusalem, the only journalist who can bare to look at it.

Think of it as: The Wire and Treme in 2099 as narrated by Hunter S. Thompson.

Big Question: Time, technology, and life all advance, but is it progress?

Just as it baffles me that there are a grand total of zero video game adaptations for television, the near complete absence of comic book adaptions for the boob tube is difficult to comprehend. Comics are, by nature, serial. Yet, despite the similarity between the genres, studios insist on compressing rich, complex comic book story arcs into an hour and a half, then crossing their fingers and hoping for a sequel. Instead, let's make one into a miniseries. Transmet and the filthy misadventures of gonzo journalist Spider Jerusalem is the perfect place to start. Patrick Stewart's production company was interested in this property and Stewart himself wrote an introduction to Vol. 5 Lonely City, but thus far nothing has happened. Let's grease the gears, shall we?

Transmetropolitan, written by Warren Ellis, follows Spider Jerusalem, a Hunter S. Thompson for the 22nd Century. After five years living in paranoid isolation on a mountain, Spider's book contracts are due. To write, he needs his fingers around the seedy, black, artificial heart of the city so that he can squeeze the tar and plaque from its arteries onto the blank pages in front of him. Spider's column is "I Hate It Here" and its popularity is directly related to Spider's level of misanthropy. He's the only writer angry enough to seek the truth and insane enough to print it. His bodyguard, Channon, and his assistant, Yelena, both as debauched and deranged as their surly boss, help Spider get into trouble and right back out of it. The show, like the comic, would follow Spider's return to the city, starting out in a disgusting apartment in the worst part of town writing about the filth and decay around him. In the comic, Spider is promoted to a new apartment as his popularity grows. The formula for the show is built right in: at the beginning of each season, Spider moves into a new apartment. In step with his rise through society, Spider's gaze moves from the filth and corruption in the gutters of the City up to the filth and corruption of the city's and country's highest offices.

Cyborgs, hybrids, uploaded nano-clouds, bowel disruptors, neuro-implants, cryonics, A.I., vat-grown meat, and a smorgasbord of transhumanist tech bursts from the background in every panel of the comic and sits at the heart of every story line. The show would be no different. *Transmet* would be an anthropological window into the City, a thriving transhuman society, the same way *The Wire* and *Treme* artfully let us into the soul of Baltimore and New Orleans.

Transmetropolitan is not a dystopia, nor a utopia, nor a caricature or a sugar coating. It is, in my mind, the fullest portrayal of what a futuristic society might actually be like — as broken and magnificent as the one in which we currently live. And with Patrick Stewart at the helm, how could it miss?

Mass Effect

Core Idea: Humanity has just survived the Contact War with Citadel Space. We've earned the right to live, now we need to earn respect, trust, and power.

Think of it as: How humans went from Battlestar Galactica to Star Trek.

Big Question: What makes humanity special? What would we add to galactic civilization?

Alright, so I couldn't very well propose a trio of SF television without getting at least one good space-based pitch in there. Again, I turn to a beloved video game with a vast SF universe — this time *Mass Effect*. In *Mass Effect*, humans are the low dogs on the totem pole, trying to earn respect among the more advanced races in Citadel Space. Facing prejudice, turf wars, and general derision wherever they go, humans have to justify themselves at every step.

But the human perspective wouldn't be the only side we would be shown. The license to touch on uncomfortable issues would allow *Mass Effect* to go where *Babylon 5* and *Star Trek* never really managed to: show the entry, rise, and constant negotiation of humanity's position within the ranks of galactic civilization and the necessary sacrifices. The central premise of *Mass Effect's* universe is that humans are late to the intergalactic game, but early in relative terms of their progress as a single civilization. Unlike the other galactic species, humans discovered intergalactic technology before achieving genetic homogeneity. Because of what is seen as extreme physical and genetic diversity and rudimentary technology, humans are considered a young, underevolved race. Having proved its worth in the First Contact war with the Turians, humans are allowed into Citadel Space, but remain a ridiculed and dismissed race. In short, *Mass Effect* could take the questions of humanity's worth raised by Q in *Star Trek* and examine them — without the camp — in rich, season-long story arcs: a perfect fit for spacefaring SF on cinematic television.

In addition to showing the human side of the struggle, *Mass Effect* could give us real, fully drawn alien protagonists. Just as *Battlestar Galactica* let us see who the Cylons were and how they thought, *Mass Effect* would show the audience why the Citadel Council doesn't welcome our species with open arms. Whole story arcs could follow how the Asari Council Envoy to Humanity both attempts to understand humans and to explain them to the Council; or the adventures of a Turian military commander forced by the council to train humans in Citadel military law. Instead of merely being told the Citadel is made up of multiple races and then, somehow when we finally see it, the members are almost entirely humanoid (I'm looking at you, Federation of Planets and Galactic Republic), there could be as many aliens as there should be. With the budget of one of the big time cable channels, space opera could finally have the makeup and SFX budget to pull off a galactic government populated primarily by non-humanoids. The diverse crew of the original games, both in terms of race and species, would give show creators precedent to cast few, if any human characters with white male actors. What better way for HBO or AMC to make their SF mark?

Three great SF shows, three awesome cable channels that pull off excellent cinematic TV, and a wide open market in which to execute them. *Caprica*, let alone Syfy, shouldn't be carrying the weight of using SF to explore the human condition on the small screen. AMC and Showtime, you want to bring HBO down a peg or two? Get the nerds on board. Looking for your next hit series, HBO? Don't be afraid of the future.

Let a Hundred Futures Bloom

A "Both/And" Survey of Transhumanist Speculation

`http://hplusmagazine.com/2009/06/15/let-`
`hundred-futures-bloom/`

Michael Garfield

Mention the word "transhumanism" to most of my friends, and they will assume you mean uploading people into a computer. Transcendence typically connotes an escape from the trappings of this world — from the frailty of our bodies, the evolutionary wiring of our primate psychologies, and our necessary adherence to physical law.

However, the more I learn about the creative flux of our universe, the more the evolutionary process appears to be not about withdrawal, but engagement – not escape, but embrace – not arriving at a final solution, but opening the scope of our questions. Any valid map of history is fractal — evermore complex, always shifting to expose unexplored terrain.

This is why I find it is laughable when we try to arrive at a common vision of the future. For the most part, we still operate on "either/or" software, but we live in a "both/and" universe that seems willing to try anything at least once.

"Transhuman" and "posthuman" are less specific classifications than catch-alls for whatever we deem beyond what we are now... and that is *a lot.*

So when I am in the mood for some armchair futurism, I like to remember the old Chinese adage: "Let a hundred flowers bloom." Why do we think it will be one way or the other? The future arrives by many roads. Courtesy of some of science fiction's finest speculative minds, here are a few of my favorites:

By Elective Surgery & Genetic Engineering

In Greg Egan's novel *Distress*, a journalist surveying the gray areas of bioethics interviews an elective autistic — a man who opted to have regions of his brain removed in order to tune out of the emotional spectrum and into the deep synesthetic-associative brilliance of savants. Certainly, most people consider choice a core trait of humanity... But when a person chooses to remove that which many consider indispensable human hardware, is he now more "pre-" than "post-?" Even today, we augment ourselves with artificial limbs and organs (while hastily amputating entire regions of a complex and poorly-understood bio-electric system); and extend our senses and memories with distributed electronic networks (thus increasing our dependence on external infrastructure for what many scientists argue are universal, if mysterious, capacities of "wild-type" *Homo sapiens*). It all begs the question: are our modifications rendering us more or less than human? Or will this distinction lose its meaning, in a world that challenges our ability to define what "human" even means?

Just a few pages later in *Distress*, the billionaire owner of a global biotech firm replaces all of his nucleotides with synthetic base pairs as a defense against all known pathogens. Looks human, smells human...but he has spliced himself out of the Kingdom Animalia entirely, forming an unprecedented genetic lineage.

In both cases, we seem bound to shuffle sideways — six of one, half a dozen of the other.

By Involutionary Implosion

In the 1980s, Greg Bear explored an early version of "computronium" — matter optimized for information-processing – in *Blood Music*, the story of a biologist who hacks individual human lymphocytes to compute as fast as an entire brain. When he becomes contaminated by the experiment, his own body transforms into a city of sentient beings, each as smart as himself. Eventually, they download his whole self into one of their own — paradoxically running a copy of the entire organism on one of its constituent parts. From there things only get stranger, as the lymphocytes turn to investigate levels of reality too small for macro-humans to observe.

Scenarios such as this are natural extrapolations of Moore's Law, that now-famous bit about computers regularly halving in size and price. And Moore's Law is just one example of a larger evolutionary trend: for example, functions once distributed between every member of primitive tribes (the regulatory processes of the social ego, or the formation of a moral code) are now typically internalized and processed by every adult in the modern city. Just as we now recognize the Greek Gods as embodied archetypes correlated with neural subroutines, the redistributive gathering of intelligence from environment to "individual" seems likely to transform the body into a much smarter three cubic feet of flesh than the one we are accustomed to.

By Nano-Hacking

Then again, there might be systemic constraints to just how far tech will take us. Charles Stross' *Glasshouse* offers a rare perspective on the possible consequences of nanotechnology: once we all rely on computers to back ourselves up and store ourselves for interstellar transit, those computers become the targets for a new level of informational warfare. In a world where people can be rebuilt at whim, murder is effectively obsolete. No one can be killed, but everyone is at constant risk of being *hacked*. Suddenly you wake up working for the enemy, and loving it. Selective memory erasure programs saturate the network and prevent any further development from crossing communities and achieving universality. History is routinely wiped, so no new wisdom can accrue. Once again, humanity is splintered into countless isolated physical and mental regions, and some of them respond by choosing to eschew high technology entirely, living and dying on the clock of some long-forgotten world.

In other words, what we normally imagine as a linear continuum might instead be a *wave* of progress that ebbs and flows, a cycle of Light and Dark Ages distributed capriciously through space-time.

By Hyperdimensional Intervention

The idea that humankind will be "initiated" into a new and higher mode of being by some other race of transcendental entities has been circulating for thousands of years. Perhaps there is a common trajectory for the development of sentient species, and we receive intermittent, minimally-intrusive guidance by those who came before us. It is an idea that has certainly found its way into common sci-fi discourse — be it through Arthur C. Clarke's *2001* or Stephen Baxter's *Manifold*. Were we to take seriously the growing ranks of exopoliticians, exobiologists, and exolinguists, this in fact is happening. Descartes was given his famous plane — practically the emblem of rational modernity — by an angelic vision. Francis Crick (co-discoverer of the double helix) and Carey Mullis (pioneer of the Polymerase Chain Reaction) both admitted to interfacing with LSD when their Nobel Prizewinning finds came to them. Crop circles form overnight in muddy fields with no footprints, bearing strange radiation signatures and seeming to encrypt dense information about the structure of the quantum vacuum and the movement of celestial bodies. This pattern is almost universal among species-changing creative eruptions (or are they irruptions?) throughout history; even Moses had his burning bush. In every instance, these revelations drew our species closer to what we might call transhuman. We're "getting the message," but who is doing the talking?

It all begs the question: are our modifications rendering us more or less than human?

By Natural Quantum Evolution

One option in particular seems to get short shrift by a community that tends to believe we will lift ourselves up into a posthuman order by our own bootstraps… but if the future even modestly resembles the past, then we cannot neglect the possibility that nature will do the heavy lifting *for* us. recent research at UC Berkeley and Washington University has demonstrated that photosynthesis is 95% efficient because it uses quantum computation to *retroactively* decide upon the best possible electron paths. Johnjoe McFadden at the University of Surrey has suggested that this very same process may have been how life emerged in the first place, and other scientists have noted similar, strangely intelligent mutation responses in lab cultures. Egan's novel *Teranesia* runs with this new model of "smart evolution," suggesting that we may see posthumanity spontaneously self-organize out of the quantum superposition of all possible futures — as if good ideas reach backward in time to organize their necessary histories. Given the uncanny prescience of some sci-fi speculation, this might not be too far from the truth.

All of The Above

As our options increase, humanity — and whatever else might call us their ancestors — will probably continue to take every form available: flesh, metal, and software; post-linguistic and pre-linguistic; evolution by self-mastery and deus ex machina. If it *can* happen, it probably will. This is the world in which we live, and every step we take into the future makes that increasingly, painfully obvious. Transhumanism, as best as I can define it, is the story of "and."

The Reluctant Transhumanist

SF Writer Charlie Stross keeps his options open

http://hplusmagazine.com/articles/artentertainm
ent/reluctant-transhumanist

R.U. Sirius & Paul McEnery

Singularity, 2032: God springs out of a computer to rapture the human race. An enchanted locket transforms a struggling business journalist into a medieval princess. The math-magicians of British Intelligence calculate demons back into the dark. And solar-scale computation just uploads us all into the happy ever after.

Stripped to the high concept, these visions from Charlie Stross are prime geek comfort food. But don't be fooled. Stross' stories turn on you, changing up into a vicious scrutiny of raw power and the information economy.

The "God" of *Singularity Sky* is really just an Artificial Intelligence, manipulating us all merely to beat the alien competition. *The Merchant Princes* (from a series of novels by Stross) are just as rapacious as anything on Wall Street, and a downstream parallel universe is just another market to exploit. *The Atrocity Archives* gives us a gutpunch full of paranoia — on the far side of hacking and counterhacking lurks an unspeakable chaos. And for all our engineering genius, *Accelerando*'s paradise is won at the cost of planetary destruction, with humanity cul-de-sac'd as our future heads off into the stars without us.

For his latest novel, *Halting State* (released in June 2008), Stross savages the fantasy worlds we escape into for fun and profit and invites us to peek underneath the surfaces as our chattering gadgets dress up reality with virtual sword-and-sorcery games, all underwritten by oh-so-creative financial instruments.

All of Stross's highly connective pipe-dream superstructures are wide open to the one geopolitical prick that will pop them all like the balloon animals they are. Be warned. Take care of the bottom line, or your second life will cost you the life that counts.

It's no surprise that Stross is a highly controversial figure within Transhumanist circles – loved by some for his dense-with-high-concepts takes on themes dear to the movement, loathed by others for what they see as a facile treatment of both ideas and characters. But one thing is certain –- Mr. Stross is one SF writer who pays close attention to the entire plethora of post-humanizing changes that are coming on fast. As a satirist, he might be characterized as our Vonnegut, lampooning memetic subcultures that most people don't even know exist.

H+: With biotech, infotech, cognitive science, AI, and so many other sciences and technologies impacting the human situation, it seems that most social and political discourse remains back in the 20th century at best. You talk sometimes about being a post-cyberpunk person. How do you deal with the continued presence of so many pre-cyberpunk people?

Charlie Stross: As William Gibson noted, "the future is already here: it's just unevenly distributed." Most people run on the normative assumption that life tomorrow will be similar to life today, and don't think about the future much. And I'm not going to criticize them for doing so; for 99.9% of the life of our species this has been the case, barring disasters such as plague, war, and famine. It's a good strategy, and periods when it is ignored (such as the millennial ferment that swept Europe around 990 A.D. and didn't die down until 1020 A.D.) tend to be bad times to live.

Unfortunately, for about the past 200 years -- that's about 0.1% of H. sapiens' life span as a species – that strategy has been fundamentally broken. We've been going through a period of massive technological, scientific, and ideological change, and it has invalidated the old rule set. But even so, at a day-to-day level, or month-to-month, things don't change so much. So most people tend to ignore the overall shape of change until it's impossible to ignore. Then they try to apply the old rules to new media or technologies, make a hopeless mess of things, and start on a slow and painful learning process. It's been quite interesting to watch the slow progress toward an international consensus on certain aspects of Internet culture, for example. In that context, I suspect the mainstream is only a decade or so behind the cutting edge: the debates over spam and intellectual property that the geeks were having in the early 1990s are now mainstream. (Of course, a decade feels like an eternity when you're up close and personal with it.)

H+: Remaining on the cyberpunk tip for a moment, Gibson's *Neuromancer* (the whole trilogy, really) popularized a trendy subculture that impacted on both entertainment and actual technology. Do you think that *Accelerando* could have that effect? Do you see yourself as a popularizer of memes that are just taking root?

CS: Naah.

A chunk of *Accelerando* was extracted in raw juicy nuggets from my time on the extropians mailing list in the early to mid-nineties; another chunk came out of my time in the belly of a dot-com's programming team in the late nineties. I wanted to get my head around the sense of temporal compression that was prevalent in the dotcom era, of the equivalent of years flickering past in months. But it's too dense for the mainstream. As we've already noticed, a lot – probably the majority – of people aren't interested in change; in fact, they find it frightening. And *Accelerando* compressed so many ideas into such a small space (I think there's about 0.5 to 1 novel's worth of ideas per chapter in each of its nine chapters) that it's actively hostile to most readers.

Some people love it, those who're already into that particular type of dense fiction-of-ideas, but many, even seasoned SF readers, just turn away.

I would like to hope that I've gone some way toward changing the terrain within the SF genre itself, though. Robert Bradbury's concept of the Matrioshka Brain (or Jupiter Brain, in earlier iterations) is one of the most marvelous SF concepts I've run across in a long time, and not trivially easy to refute. I wanted to get past the then-prevalent idea that you couldn't write about a Vingean singularity – it's difficult, but we've got tools for thinking about these things. And I got the idea of computronium into common enough parlance that Rudy Rucker recently took a potshot at it, implying that it's part of the universe of discourse in my field.

H+: I'm curious about the Economics 2.0 idea that is featured in *Accelerando*. What do you think about economic systems in a presumably posthuman world? Do any of the theories — free market, Marxist, and so forth — that have guided those who ideologize these things continue to make sense after replicators and the like?

CS: In a nutshell, about Economics 2.0: economics is the study of the allocation of resources between human beings under conditions of scarcity (that is, where resources are not sufficient to meet maximal demand by all people simultaneously). Resource allocation relies on information distribution -- for example, price signals are used to indicate demand (in a capitalist economic system). In turn, economic interactions within, for example, a market environment hinge on how the actors within the economic system use their information about each other's desires and needs.

To get a little less nose-bleedingly abstract: say I am crawling through a desert and dying of thirst, and you happen to have the only bottled water concession within a hundred miles. How much is your water worth? In the middle of a crowded city with drinking fountains every five yards and competing suppliers, it's worth a buck a bottle. But in the middle of a desert, to someone who's dying of thirst, its value is nearly infinite. You can model my circumstances and my likely (dying-of-thirst) reaction to a change in your asking price and decide to hike your price to reflect demand. You can do this because you have a theory of mind, and can model my internal state, and determine that when dying of thirst, my demand for water will be much higher than normal. And this is where information processing comes into economic interactions.

What kind of information processing can vastly smarter-than-human entities do when engaging in economic interactions? In *Accelerando* I hypothesized that if you can come up with entities with a much stronger theory of mind than regular humans possess, then their ability to model consumer/ supplier interactions will be much deeper and more efficient than anything humans can do. And so, humans will be at a profound disadvantage in trying to engage in economic interactions with such entities. They'll be participating in economic exchanges that we simply can't compete effectively with because we lack the information processing power to correctly evaluate their price signals (or other information disclosures). Hence Economics 2.0 – a system that you needed to be brighter – than human to participate in, but that results in better resource allocation than conventional economic systems are capable of.

H+: What do you think about transhumanism and singulari-tarianism as movements? Are these goals to be attained or just a likely projection of technologies into the future that we should be aware of?

CS: My friend Ken MacLeod has a rather disparaging term for the singularity; he calls it "The Rapture of the Nerds."

This isn't a comment on the probability of such an event occurring, per se, so much as it's a social observation on the type of personality that's attracted to the idea of leaving the decay-prone meatbody behind and uploading itself into AI heaven. There's a visible correlation between this sort of personality and the more socially dysfunctional libertarians (who are also convinced that if the brakes on capitalism were off, they'd somehow be teleported to the apex of the food chain in place of the current top predators).

Both ideologies are symptomatic of a desire for simple but revolutionary solutions to the perceived problems of the present, without any clear understanding of what those problems are or where they arise from. (In the case of the libertarians, they mostly don't understand how the current system came about, or that the reason we don't live in a minarchist night-watchman state is because it was tried in the 18th and 19th centuries, and it didn't work very well. In the case of the AI-rapture folks, I suspect there's a big dose of Christian millennialism (of the sort that struck around 990–1010 A.D., and again in the past decade) that, because they're predisposed to a less superstitious, more technophillic world-view, they displace onto a quasiscientific rationale.

Mind uploading would be a fine thing, but I'm not convinced what you'd get at the end of it would be even remotely human. (Me, I'd rather deal with the defects of the meat machine by fixing them – I'd be very happy with cures for senescence, cardiovascular disease, cancer, and the other nasty failure modes to which we are prone, with limb regeneration and tissue engineering and unlimited life prolongation.) But then, I'm growing old and cynical. Back in the eighties I wanted to be the first guy on my block to get a direct-interface jack in his skull. These days, I'd rather have a firewall.

H+: You said "I'd be very happy with cures for senescence, cardiovascular disease, cancer, and the other nasty failure modes to which we are prone, with limb regeneration, and tissue engineering and unlimited life prolongation." It seems to me that this still puts you in the Transhumanist camp. Would you agree?

CS: To the extent that I don't believe the human condition is immutable and constant then yes, I'm a Transhumanist. If the human condition was immutable, we'd still be living in caves. (And I have a very dim view of those ideologies and religions that insist that we shouldn't seek to improve our lot.)

H+: Earlier on, you referred to the Matrioshka brain. Can you say a bit more about that and why you find it an appealing or, perhaps, realistic concept?

CS: As I said, the credit for the concept belongs to Robert Bradbury, who refined it further from discussions by Eliezer Yudkowsky and others in the mid-nineties, in turn based on speculation by Freeman Dyson going back as far as the 1960s.

Dyson first opened the can of worms by suggesting that we could make better use of the matter of the solar system by structuring it as free-flying solar collectors and habitats in variously inclined but non-intersecting orbits, which would trap the entire solar radiation output and give us access to mind-numbingly vast amounts of energy and inhabitable space.

The extropians took the idea one step further, with the idea of computronium — the densest conceivable form of matter structured to maximize computation. What amount of thinking can you get done by building a Dyson sphere, optimized to support computation rather than biological life? Bradbury suggested building multiple concentric spheres of free-flying compute nodes, each shell feeding off the waste heat of the next layer in. Some estimates of the computing power of such a Matrioshka Brain (named after the nested Russian dolls) suggest that it would be roughly as far beyond us – the entire human species – as we are beyond a single nematode worm.

If the idea of procedural artificial intelligence holds water, it's possible that a Matrioshka Brain (or something like it) is going to turn out to be the end state of any tool-using civilization: after all, the bulk of the mass of which our planet is composed is of no use to us whatsoever (other than insofar as it makes a dent in spacetime for us to stick to), never mind the rest of the solar system...

H+: Moving on, your latest novel, *Halting State* is all about different levels of reality. LARPs and Second Life, office politics, the "mammalian overlay" of sexual seduction, financial instruments: they're all artificial realities, one layer on top of each other, and all interacting. It's sort of like what we used to think of as a spiritual realm, but it's right here running on TCP/IP. It used to be only shamans and schizophrenics who had these sorts of visions, but now, if you're wearing the special specs, we all get to share this world that's haunted by imaginary beings. I think of Arthur C. Clarke's notion that a sufficiently advanced technology is indistinguishable from magic. Do you think the areas and powers that we're opening up will change us?

CS: What makes you think it's about us?

We're human 1.0. We're not going there. Or we may go down that road, but the things that arrive at the other end won't be us. (They might remember having started out as us, but I'm not betting on it.)

H+: There's a nasty little idea buried in *Halting State*, I think. Like... If you think things are bad when people get their ideas about reality from TV, wait until our imaginations are completely colonized, surveilled and programmed. Our hero bleakly opines, that this is the reason for the Fermi Paradox. There are no signs of alien life because you get so far and then vanish up your own artificial reality. Have I got that right? And is that a prediction?

CS: I try not to make predictions -- but I see that one as a distinct possibility (and indeed, as yet another solution to the Fermi Paradox).

Was Michael Jackson a Transhumanist?

http://hplusmagazine.com/editors-blog/was-michael-jackson-transhumanist

R.U. Sirius

The rumor — whether true or not — that Michael Jackson missed his cryonics appointment, but instead arrangements are being made to have him plastinated by the ever-controversial Gunther von Hagens would not surprise anybody, if true. Rather, one is — to some degree — surprised that Jacko did not arrange for the greatest return engagement in entertainment history.

On the other hand, the corporeal world did not treat him altogether kindly. Sure, he had riches to spend (his own, and then other people's) and he was positioned to give dramatic expression to his ego in extraordinary ways. Indeed, after the phenomenal success of *Thriller,* he could declare himself the King of Pop and more or less get away with it. (Perhaps he should have had himself entombed instead like Tutankhamen. Or given his use of Stalinist imagery to promote *HIStory,* he might have chosen to preserve himself in a glass case along the lines of Mao and Lenin). Whatever. Jackson went through serious trials and tribulations and received a lot of abuse for being "weird."

The popular story is that Jackson was loved and respected for his "innovative" music — his song and dance act, and his videos, but he alienated a good portion of his audience by doing odd stuff. Some of us may prefer to flip that around.

While Jackson was a marvelous singer and performer, and he had some fine and funky musical moments, his musical oeuvre was entirely conventional, as were most of his videos. Indeed, Jackson — and the lyrical and musical sensibilities that he represented — were basically reactionary. It was a step backwards from the musical innovations popularized by The Beatles and others (including others like George Clinton and Sly Stone, in the funk genre). A "top 40" musical culture that could incorporate the lyrical ideas of a Dylan or a Bowie or a Curtis Mayfield or a Kurt Cobain has largely been displaced by the empty entertainment tropes represented by *American Idol,* and Jackson is the idol of "Idol."

Then again, there's another way of looking at that story. In the days when rock was "important" — for every David Bowie or John Lennon, there were hundreds like Uriah Heep, Emerson Lake and Palmer, Billy Joel... "Poets" who wanted to express themselves via rock and roll but who maybe should have known better. The "Disco Sucks" movement may have been partly a reaction to the dumbing down of rock lyrics (or at least an abandonment of its quasi-adolescent rebellious themes), but it was also a reaction against disco's mixed-race, mixed-sexual preference, multi-gendered, urban party uprising.

Jackson's fifth (and best) solo album, *Off The Wall,* was released in 1979, during the disco era. While he wasn't a disco artist in the sense that Gloria Gaynor was, it *was* dance music and it went into heavy rotation at Studio 54 and no doubt wherever club goers gathered around the globe.

Jackson emerged as a solo artist into a rock culture that still had a hint of androgyny (from the '70s) and in which the word "freak" was a compliment rather than an insult (from the '60s). Nonconforming celebrities were cheered on, as opposed to being snarked to death.

While for the most part, Jackson paid attention throughout his career to no one other than himself, some of that zeitgeist may have rubbed off on him. Jackson was a '"permissive" '70s person. And had he been an '80s person, he may have been altogether more constrained in his pursuits.

So Jackson, in his own way, "freaked freely"... but was he a transhumanist?

In an interview in our first issue (run elsewhere in this book under the title "Botox Parties, Michael Jackson, and the Disillusioned Transhumanist"), Christopher Dewdney concentrates on Jackson's defiance of his biological limitations in terms of his transformation of his inherited features — his cosmetic surgery and his change of skin color (he might also have included his cross-genderish style of dress and makeup).

In *More Than Human,* Ramez Naam predicted that during the next decade, we will be able to genetically alter our skin colors (and this would not necessarily be limited to the usual boring choices. Start *being* purple?). This would certainly be a radical transhuman shift, but changing his features is not the only way Jackson gave expression to a dissatisfaction with "natural" life as a given, and showed his interest in mutation and expanded possibilities — and dabbled (at the very least) in transhumanist tropes. We can also include his use of a hyperbaric oxygen chamber — according to rumors, in hopes that it was a key toward longevity. Foolish, certainly, but another indication that Jackson was reaching for something other. Jackson's fascination with mutation — for better and for worse — was also manifested in his purchase of John Merrick's (the "elephant man") skull. Hey, he even prefigured the contemporary fascination with Zombies with his most popular video for *Thriller.*

Jackson also lived in a sort-of virtual reality. Having reached his adulthood (sort of) in an age before VR, he used his wealth to build his own alternative reality in the only way he could — physically —as manifested most particularly on his Neverland Ranch. There, Jackson "programmed" his own version of a J.M. Barrie Peter Pan world, and populated it with children (and firewalls, otherwise known as body guards). He proceeded to marry a virtual reality — the daughter of the King — and, later, via surrogates — he took on his own virtual children to raise up. As a child star, Jackson was in some ways predestined to live a virtual life, and the abuse he experienced at the hands of his father guaranteed that he would pursue fantasy (and the fantastic).

Last but not least, there is Jackson's neotony. Quoting Wikipedia: "Neoteny (pronounced /niːˈɒtɪniː/), also called juvenilization, is the retention, by adults in a species, of traits previously seen only in juveniles (a kind of paedomorphosis), and is a subject studied in the field of developmental biology. In neoteny, the physiological (or somatic) development of an animal or organism is slowed or delayed (alternatively, seen as a dilation of biological time)." Neotony in a species can be an evolutionary trait — a sign that a species is going to mutate within generations. In the 1970s, Timothy Leary — himself an early transhumanist – popularized the notion that the refusal of boomer and post-boomer generations to embrace traditional "terminal" adulthood signified a coming evolution in the human species.

Leary's notion of paedomorphosis was that generations would remain in a sort of adolescent phase up until the birth of the posthuman. But Jackson never got that far. The private "virtual" M.J. was a child — a pre-teen — and that's where the Jackson story went sour. If Jackson diddled any pre-teens, this is — of course — reprehensible. Even those of us at the edge of this dynamic period in transitional humanity must recognize some boundaries. Pathology may precede potential in this odd and challenging transition phase moving toward an incomprehensible future, but some virtual fantasies must remain virtual, if at all.

Of course, M.J. was never convicted of sexual acts with children. But he probably died as the result of the emotional strain brought on by his trials, his public shunning, and his period of self-exile, or from the prescription painkillers he took to dull that pain.

In the end, Jackson's life reads as tragedy, but context is everything. Given the opportunity, most of us would probably take some of the drama along with some of the glory, the adventure, the money, the indulgences, the ego strokes.... I mean, it was a pretty rich life... In all senses of the word.

Michael Jackson is obviously not an example of transhumanism to be followed. But he is a signpost on the road to posthumanity.

I believe the future will study him from that perspective, and in some odd way, it will learn from his many mistakes.

Gene Genies: BIO

The Great Designer Baby Controversy of '09: The Fertility Institute Backs Away From Making History

http://hplusmagazine.com/articles/bio/great-designer-baby-controversy-%E2%80%9909

Michael Anissimov

You may not know it, but gender selection based on pre-implantation genetic diagnosis (PGD) has been available to paying couples since at least 2001. One of the world leaders in providing this service is the Fertility Institutes, with branches in Los Angeles, New York, and Guadalajara in Mexico. According to their website, they've had over 3,800 cases of gender selection with a 100% success rate. Besides offering gender selection, they screen embryos for genetic defects such as breast cancer, cystic fibrosis, and over 70 other diseases. The Institutes are directed by Dr. Jeff Steinberg, a pioneer of IVF (in vitro fertilization) in the 1970s, and a successful scientist-businessman today.

In early February, the Fertility Institutes created enormous controversy by announcing that they planned to offer PGD services allowing for the selection of eye and hair color for children. Steinberg was quoted by the BBC as saying, "I would not say this is a dangerous road. It's an uncharted road." As a scientist experienced in PGD/IVF techniques, Steinberg was aware that the technology to select physical traits in humans has been available for years, but no one would touch it. "It's time for everyone to pull their heads out of the sand," Steinberg said. Transhumanists and other fans of procreative freedom were excited by the news.

The backlash was widespread. Quoted in the *New York Daily News* on February 23, the Pope himself condemned the "obsessive search for the perfect child." The pontiff complained, "A new mentality is creeping in that tends to justify a different consideration of life and personal dignity." The Roman Catholic Church objects to all applications of PGD because they invariably involve the destruction of blastocysts.

On his blog Secondhand Smoke, conservative bioethicist Wesley J. Smith, who has co-authored four books with Ralph Nader, wrote, "We are constantly told that the right of a woman to reproduce is absolute, including getting pregnant, aborting if the pregnancy is ever unwanted, and now, genetically engineering progeny to order. But no 'right' is absolute. The time has long since passed to put some regulatory controls over the wild, wild west of IVF."

On February 28, Steinberg continued to defend his approach by telling the *Sunday Telegraph,* "I understand the trepidation and concerns, but we cannot escape the fact that science is moving forward. If I have to get smacked around by people who think it is inappropriate, then I'm willing to live with that."

Then, all of a sudden, on March 2, Steinberg capitulated to widespread criticism. A press release on the Fertility Institutes web site read, "In response to feedback received related to our plans to introduce preimplantation genetic prediction of eye pigmentation, an internal, self-regulatory decision has been made to proceed no further with this project." *Gattaca* was averted.

The public debate about selecting traits like eye and hair color for newborns is a continuation of a debate that has gone on for at least two decades — the debate about PGD-based gender selection, a technique that is easier than trait selection and has already been done thousands of times. Back in 1990, pre-implantation genetic diagnosis of any type was banned in Germany by the Embryo Protection Act. In 2003, the UK banned using PGD for gender selection, following a year-long public consultation in which about 80% of those polled were against the procedure. India and China have banned the procedure, despite the widespread practice of infanticide when babies of an undesired gender, usually female, are born to disappointed parents. Gender selection still occurs, albeit violently.

More recently, a January 2009 study by researchers at NYU Langone Medical Center found that an overwhelming 75% of parents would be in favor of trait selection using PGD – as long as that trait is the absence of mental retardation. A further 54% would screen their embryos for deafness, 56% for blindness, 52% for a propensity to heart disease, and 51% for a propensity to cancer. Only 10% would be willing to select embryos for better athletic ability, and 12.6% would select for greater intelligence. 52.2% of respondents said that there were no conditions for which genetic testing should never be offered, indicating widespread support for PGD – as long as it's for averting disease and not engineering human enhancement.

James Hughes said: "The term 'designer babies' is an insult to parents, because it basically says parents don't have their kids' best interests at heart."

Trait selection using PGD is too new – and unproven – for there to be regulatory laws in most developed countries. But many fighters in the battle for or against PGD for trait selection and genetic disease screening believe that today is the decision point that will set the precedent for future regulation (or lack thereof) in the area. On May 21, 2008, the US Congress passed the Genetic Information Non-Discrimination Act. According to the statement of Administration Policy associated with the Act, it "prohibit[s] group health plans and health insurers from denying coverage to a healthy individual or charging that person higher premiums based solely on a genetic predisposition to developing a disease in the future. The legislation also would bar employers from using individuals' genetic information when making hiring, firing, job placement, or promotion decisions. The Administration appreciates that the House bill clarifies that the bill's protections cover unborn children."

In the oft-cited movie *Gattaca,* a nongenetically selected man with a heart problem in a trait-selected world must hide his status through the course of his ambition to become an astronaut. Theoretically, the 2008 law would make this type of discrimination illegal, at least in the United States. But what about *Gattaca?* The film was invoked so frequently in negative responses to the Fertility Institutes' announcement that it is hard to find a comment thread on the topic that doesn't mention it. In his 2004 book *Citizen Cyborg,* Dr. James Hughes, a transhumanist bioethicist and director of the Institute for Ethics and Emerging Technologies, pointed out a few quibbles with the movie:

1. Astronaut-training programs are entirely justified in attempting to screen out people with heart problems for safety reasons.

2. In the United States, people are already discriminated against by insurance companies on the basis of their propensities to disease despite the fact that genetic enhancement is not yet available.

3. Rather than banning genetic testing or genetic enhancement, society needs genetic information privacy laws that allow justified forms of genetic testing and data aggregation, but forbid those that are judged to result in genetic discrimination (such as the previously mentioned U.S. Genetic Information Nondiscrimination Act). Citizens should then be able to make a complaint to the appropriate authority if they believe they have been discriminated against because of their genotype.

Those on the other side of the divide are numerous. At a 2008 meeting of the American Society of Human Genetics, William Kearns, a leading medical geneticist, when prompted about trait selection, said "I'm totally against this. My goal is to screen embryos to help couples have healthy babies free of genetic diseases. Traits are not diseases." Mark Hughes, the head of the Genesis Genetics Institute in Detroit, has called the practice "ridiculous and irresponsible." More bluntly, George Annas, a bioethicist with Boston University, has said "modern genetics is eugenics," while on a visit to the Holocaust Museum in Washington, DC.

The falling costs of gene sequencing is enabling PGD trait selection and lowering the barrier to entry. In the last few years, the cost of sequencing a base pair has fallen so low that even the optimists have been surprised. The first human genome that was sequenced, by the federally financed Human Genome Project in 2003, cost a few hundred million dollars. In 2007, sequencing James Watson's genome cost about $2 million. In March 2008, Applied Biosystems, based in California, sequenced a genome in two weeks for $50,000. In June 2010, the genomics company Illumina announced that it would be offering full genome sequencing for just $19,500, as low as $14,500 when groups of five or more people are referred by the same physician, and just $9,500 when the patient's doctor certifies that the sequencing could lead to better treatment of the patient's disease or condition. Over a dozen companies are aiming to bring the price of full genome sequencing down to less than $1,000 before 2013

The requisite technologies for trait selection are on the way, but the battle lines have not yet been entirely drawn. Prompted by a *Wall Street Journal* article on the Fertility Institutes and trait selection, Kathryn Hinsch of the Women's Bioethics Project argued that thinking about the issue carefully is important, and refrained from taking a hard stance on either side. She said that trait selection should be considered because, "1) It's a hive of ethical issues, 2) The technology isn't here yet, 3) We all have a stake in the issue, and 4) Questions raised go beyond designer babies." According to Hinsch, the key questions that need to be addressed are: "Should we ban it? Should we regulate the technology to allow only certain applications? Should we promote the widespread use of this technology?"

The advocates of trait selection using PGD, at least in the Western world, appear to be small in number. But as the NYU Langone Medical Center survey showed, there are at least a few. On his blog Sentient Developments, George Dvorsky, a prominent transhumanist bioethicist, pointed out that "some demand is still demand." Commenting on the survey, Dvorsky said, "An anti-enhancement bias is most certainly embedded in our society. It's very likely that many of the respondents were answering the survey in accordance to their social conditioning and what they thought was expected of them from an 'ethical' perspective." Supporting the idea of trait selection, Dvorsky wrote, "What we're talking about here is endowing our children with all the tools we can give them so that they may live an enriched, open-ended and fulfilling life. By denying them these benefits we are closing doors and potentially reducing the quality of their lives." Another advocate of cautious trait selection is Ramez Naam, author of the 2005 book *More Than Human.* In a chapter on genetic engineering, he writes, "A regulatory regime consistent with family choice would focus on safety, education, and equality rather than prohibition." Looking past the immediate future, Naam also writes, "Ultimately, whatever choices we make for our children will be subject to change, at their choice, when they reach adulthood. In the coming years, pharmaceuticals, adult gene therapy, and the integration of computers into the brain will give people far more control over their own minds and bodies than we enjoy today."

In a March 9, 2009 *WIRED* online interview, James Hughes registered support for trait selection, and also railed against the "designer baby" terminology altogether. Responding to the future of trait selection, he said, "It's inevitable, in the broad context of freedom and choice. And the term 'designer babies' is an insult to parents, because it basically says parents don't have their kids' best interests at heart." He said, "If I've got a dozen embryos I could implant, and the ones I want to implant are the green-eyed ones, or the blond-haired ones, that's an extension of choices we think are perfectly acceptable — and restricting them a violation of our procreative autonomy."

PGD and other reproductive technologies are commonly rejected as "unnatural". The transhumanists and technoprogressive response is summarized well in the Transhumanist FAQ, which says, "In many particular cases, of course, there are sound practical reasons for relying on 'natural.'

processes. The point is that we cannot decide whether something is good or bad simply by asking whether it is natural or not. Some natural things are bad, such as starvation, polio, and being eaten alive by intestinal parasites. Some artificial things are bad, such as DDTpoisoning, car accidents, and nuclear war."

The legal and ethical future of trait selection based on PGD is still unknown. What is known is that parents will always want the best for their children. When push comes to shove, they will probably take advantage of whatever technologies are available that will give them the best lives possible.

Adventures in Synthetic Biology: Interview with Stanford's Drew Endy

http://hplusmagazine.com/articles/bio/adventure
s-synthetic-biology

Surfdaddy Orca

In the first few panels of Drew Endy's "Adventures in Synthetic Biology" comics you see a young student with laboratory goggles grabbing Buddy-the-Lifeform. His instructor takes him inside a cell to view the organism's genome, "the master program that's running the cell." The young student remarks, "So this is what we change to reprogram this critter? LOOKS EASY!"

Dr. Endy's innovative comics attempt to make the nascent field of synthetic biology "look easy" to both student and layman alike. Synthetic biology combines science and engineering in order to design and build (through DNA synthesis) novel biological functions and systems. Endy's comics made the cover of *Nature* in 2005.

Dr. Endy's day job is as an associate professor at Stanford University's Department of Bioengineering. He's helping to build a library of standardized biological components for use in genetic research. While his research interests are varied, they focus on engineering integrated biological systems.

As reported in *Nature*, engineered biological systems have been used to manipulate information, construct materials, process chemicals, produce energy, provide food, and help maintain or enhance human health and our environment. Dr. Endy envisions an open technology platform for engineered biological systems that is component-based, standardized and reliable.

Drew Endy earned degrees in civil, environmental, and biochemical engineering at Lehigh and Dartmouth. He did postdoctoral studies in genetics and microbiology at UT Austin and UW Madison. From 1998 through 2001 he helped to start the Molecular Sciences Institute, an independent not-for-profit biological research lab in Berkeley, California.

He joined the MIT faculty in 2004 where he co-founded the MIT Synthetic Biology working group and the Registry of Standard Biological Parts. He also organized the First International Conference on Synthetic Biology. With colleagues, he taught the 2003 and 2004 MIT Synthetic Biology labs that led to the organization of iGEM, the International Genetically Engineered Machine competition. Teams of students from schools around the world compete regularly in iGEM.

In 2004, Dr. Endy co-founded Codon Devices, Inc., a venture-funded startup that worked to develop next-generation DNA synthesis technology. A year later he co-founded the BioBricks Foundation (BBF), a not-for-profit organization founded by engineers and scientists from MIT, Harvard, and UCSF with significant experience in both non-profit and commercial biotechnology research. BBF encourages the development and responsible use of technologies based on BioBrick™ standard DNA parts that encode basic biological functions.

H+ contacted Dr. Endy at his Stanford University office.

H+: My understanding is that the goal of synthetic biology is to build a biological machine — or modify an existing organism — using standard parts, much like a computer or electrical engineer might design and build a computer using off-the-shelf microchips and circuit boards. Are we close to doing that now?

DREW ENDY: I don't know that we're close to doing it at all. I think what we're trying to do is get better at engineering biology. We're looking to past examples, where, in other types of engineering, people have developed improved capabilities for engineering different types of material such as metals, silicon, and so forth. We're adapting those lessons to the substrate of biology and seeing which of them might be useful. I think it's an area of research — which is important to emphasize — because what it means practically is that it's not at all obvious that the lessons from computer engineering, electrical engineering, mechanical engineering, or civil engineering will directly translate and apply to the substrate of life. They inform us and provide points of departure. It would also be surprising if they didn't have something to add as well. So, the work of synthetic biology, which as a field in its modern form is only about five-years-old, is an opportunity to work on improving the process of designing, constructing, measuring, testing, debugging, and reworking living components — as well as living systems — to see if we can improve how we work with them and partner with them to do different things.

I don't think it's at all imminent that we are going to be designing and building new life forms from whole cloth. What's imminent are many opportunities to get better at the process of engineering living systems, starting from the ones that already exist, and seeing what comes next.

H+: **In your talk at First International Conference on Synthetic Biology at MIT in 2004, you outlined four new inventions that enabled today's synthetic biology: DNA synthesis, standard-ization, abstraction, and decoupling. How much have we progressed in each of these areas since the first conference?**

DE: Synthesis of DNA is based on a chemical process that was perfected in 1982. What's happened over the past five years since the 2004 conference is, from a commercial perspective, that the cost of having a gene synthesized has dropped from about $4.00 to somewhere between $0.50 and $1.00. A four-to eight-fold improvement. That's a lot, but perhaps not as much as some people might hope for — this means there's a lot of room for improvement.

In terms of standardization, there is a first generation of standard biological parts called the BioBricks parts collection out of MIT. It probably numbers about 5000 parts. It was motivated by naive allusions to the Transistor-transistor Logic Data Book that Texas Instruments and other electronics manufacturers might provide.

No one should be under any illusion that standard biological parts collection as it now exists is of the same quality or maturity as what people might be familiar with in electronics. Nevertheless, it's a fantastic success story. It's growing exponentially with the number of parts being added to the collection year-after-year. It's also become a world resource that's unparalleled and quite valuable. Students and others who are starting genetic projects today don't have to start from scratch. This allows a lot of projects to happen much more quickly than was possible five years ago.

The next step, in my mind, around parts collections is to professionalize them so that they support the worldwide diversity of folks who are identifying all sorts of new natural genetic functions that might be adapted for technology purposes. We want to support those folks with core, professional teams that are working for public benefit while making very high quality standard biological parts as best as anyone now knows how and giving them back to the world as an open technology platform for the future of cellular and genetic engineering.

H+: This begins to sound very much like Open Source software in the computer domain.

DE: Yeah. And to speak frankly, I think synthetic biology is informed by the transition in computing from AT&T and Unix circa 1971 to today. If you look at the transitions in computing over that period of time, some of the first software that AT&T wrote is word processing software for writing patent applications. There was a move to use copyright as a legal mechanism for defining and sharing innovation frameworks in software through the 1980s up to this day. For a period of time that led to a split in the software development community where some folks decided to use copyright to define proprietary software and generate significant revenue streams around that — Bill Gates and

Microsoft is a leading example. Other people decided to use copyright to enforce standards of freedom around software — that's Richard Stallman of the Free Software Foundation.

But from the 1980s to today it took about two decades to have a rich ecology of software innovation. Today you find quite a vibrant ecosystem of software that includes open technology platforms comprising Linux, mySQL, Apache as a web server, and so on.

I think the opportunity that presents itself to the biotechnology community is that we can transition to a richer innovation framework — and I mean richer in many ways — and get to open technology platforms without a whole lot of dysfunction as an intermediate stage.

Coming back then to your bigger question [about progress since the first conference], we've talked about construction and we've talked about standards. Abstraction as an idea came into existence for genetic circuits based on transcription — that is, the reading out of DNA — and it's been extraordinarily powerful as a first step, but it hasn't gone much beyond that. The PoPS (Polymerase Per Second) standard, while not yet widely understood, is quite powerful. But the reading out of DNA is just one type of cellular function. There are many others, and it would be good to see the ideas of abstraction explored against different categories of biological function.

H+: The fourth area that enabled synthetic biology is decoupling, right?

DE: Decoupling is a consequence of the success of the other areas. For example, if you have the ability to print DNA on demand, this means that one person can be a designer of DNA and another person can be a builder of DNA. That's a type of decoupling, like an architect and a contractor. However, if the designer of DNA does not have a language or a grammar that supports the programming of many genetic systems, then the decoupling between designer and builder will be useless. Where do these languages and grammars come from? They come from advances in standards and abstraction and any other idea that might apply.

I think there've been some good steps towards decoupling. We're seeing teams using DNA synthesis and acting as designers — but we still have to do a lot of work because the component sets themselves are not very mature. What the sets are calling on are not reliable objects... They're research projects basically.

H+: A number of these ideas were encapsulated in your "Adventures in Synthetic Biology" comics that made the cover of *Nature* magazine. A comic book seems like an innovative way to communicate complex ideas to a younger audience. Has it been included in educational curricula? Do you have any plans to make it into a series?

DE: I would love to take it forward as a series of comic books. To be fair, it was a lot of work as you might imagine. You're highlighting quite sharply many of our motivations around doing it. It was a project that we took forward as an educational project, meaning that we were really struggling to communicate certain ideas to our students, specifically the idea of a common carrier for genetic circuitry. In drafting the comics, what we tried to do — and this is manifested in the last chapter of the comics with the title "Common Signal Carrier" — is insert a dialog between student and teacher that would lay out on one page all the questions and puzzles that we found students to be wrestling with when we tried to explain this sort of idea. And I think we did a good job of that, to be frank.

H+: I think you did too. I was very impressed.

DE: But, if I then try to assess the impact of the comic book in teaching these ideas, I don't know if it was a success. Often it's difficult for people to imagine that they might learn engineering theory from a comic book. Maybe they don't get to the third chapter. They just read the first chapter, which is simpler and meant to explain what's happening to a person who might not know what DNA is. I'd be curious if you or any of your readers might have some insight on that. I recognize that the comic book form is very powerful. But as a pedagogical device, I'm not convinced that it is as impactful as we might have hoped.

H+: It seems like a great approach to communication, and I'm curious to know if you think that the comic idea will go anywhere as an educational device.

DE: Well, we don't have a communication strategy for it (laughs). But I think it will. I'm speaking from my own perspective. One of my colleagues at MIT extended the comic idea via a web site called biobuilder.org You'll see that he uses comic-like animation around specific aspects of synthetic biology.

H+: I see from your web site that there are quite a few projects being pursued by the Endy Lab at Stanford — genetic memory, computational modeling, electronic counter review, DNA sequence refinement, synthetic biology data transfer protocol, to name a few — can you talk a little about your current research focus?

DE: The lab at Stanford is for the students. The research at the lab is really being defined by the students as they learn to become better researchers. The meta-level framing of the lab is focused on getting much better at putting together parts to make integrated genetic systems and having those systems work as we expect. So, for example, the types of systems that synthetic biologists are building today might contain ten to twenty different genetic components. But the pieces of DNA that people can construct are capable of containing thousands of components. So researchers can put together eight million base pair fragments of DNA that can have 8,000 different genetic parts on it. But these are not novel engineered systems, they are recapitulations — we're "plagiarizing" the natural sequences. Meanwhile, we engineers can design, build, and get working 10,000, 20,000, maybe up to 60,000 base pair fragments. This means that we have an opportunity to get between 100 to 800 times better at genetic engineering.

And how are we going to get better at putting parts together? The answer is that we don't know, so this becomes a research topic in itself. One of the reasons we're interested in implementing scalable information storage systems inside cells is that the scoping of these systems is about ten times more complicated than anything anyone has been able to build today. So by targeting that sort of system, we're challenging ourselves to get better at integrating genetic components into many component systems.

All the specific projects in the lab are exploring ways to take forward the opportunity to get better at putting things together, specifically, the integration of genetic components.

H+: Your startup company, Codon Devices, closed its doors recently. Can you talk a little bit about what happened? Is the market not quite ready for synthetic biology devices?

Drew Endy: Codon Devices went to market as a gene synthesis company. It did not compete successfully as a gene synthesis company and so we shut it down and went out of business. We didn't go bankrupt, which I'm proud of in some strange way. I think the gene synthesis marketplace generally is doing quite well and has a lot of competition. This marketplace continues to grow, and there seem to be many opportunities for improvement in gene synthesis technology. People are continuing to work on it.

H+: One last question if I may. What is your vision for synthetic biology over the next five to ten years?

DE: The things I'm anxious to see over the next five years include getting the construction technologies to be much more powerful and practical. If we could drop the cost of gene synthesis by a factor of 100, that seems within striking distance. It basically requires obtaining oligonucleotides [short nucleic acid polymers, typically with twenty or fewer bases] for building genes in some reliable fashion from oligo chips [chips containing several thousand oligonucleotides]. People have demonstrated that this is possible, but no one has made it reliable as a commercial process.

A second goal for me over the next five years is to see a very professional biological parts collection, complementing what the student competitions are now producing. These collections can be used to form an open technology platform that future biotechnology can draw upon for free to more cheaply and effectively deliver solutions that address the problems and opportunities that the world presents to our civilization.

So, in summary, about a hundred-fold improvement in the cost of construction would be nice, and a professionalization of the parts collection, so that, for example, the central dogma in microorganisms like *E. coli* and yeast no longer present themselves as research questions for genetic engineers. I'm sure there will continue to be scientific research into these microorganisms, but I don't want future genetic engineers to have to debug the equivalent of a "print" statement. The other big goal for the next five years is to see if we can catalyze an open technology platform in this sector that brings industry and academia together in a new venue to do a lot more and more effectively.

In terms of the applications in the field [materials, chemicals, energy, food, human health, etc.], it's actually my job not to speculate. The whole significance of synthetic biology is that we're not trying to overdrive it with any one application. We're trying to invent and practice new technology platforms and languages and grammars that enable biotechnology quite broadly. I'd look instead to what the students in iGEM and other educational programs are exploring as leading indicators of what the future of biotechnology might hold.

Eight Ways In-Vitro Meat Will Change Our Lives

http://hplusmagazine.com/articles/bio/eight-ways-vitro-meat-will-change-our-lives

Hank Hyenas

"Future Flesh" is squatting on your plate. Are you nervous? Stab it with a fork. Sniff it. Bite! Chew, swallow. Congratulations! Relax and ruminate now because you're digesting a muscular invention that will massively impact the planet.

In-Vitro Meat – aka tank steak, sci fi sausage, petri pork, beaker bacon, Frankenburger, vat-grown veal, laboratory lamb, synthetic shmeat, trans-ham, factory filet, test tube tuna, cultured chicken, or any other moniker that can seduce the shopper's stomach – will appear in 3-10 years as a cheaper, healthier, "greener" protein that's easily manufactured in a metropolis. Its entree will be enormous; not just food-huge like curry rippling through London in the 1970's or colonized tomatoes teaming up with pasta in early 1800's Italy. No. Bigger. In-Vitro Meat will be socially transformative, like automobiles, cinema, vaccines.

H+ previously discussed In-Vitro Meat, as have numerous other publications [see references at the end of this article]. Science pundits examined its microbiological struggles in Dutch labs and at New Harvest, a Baltimore non-profit. Squeamish reporters wasted ink on its "yucky" and "unnatural" creation, while others wondered if its "vegan" or not (PETA supports it but many members complain). This article jumps past artificial tissue issues; anticipating success, I optimistically envision Eight Ways In-Vitro Meat Will Change Our Lives.

Bye-Bye Ranches

When In-Vitro Meat (IVM) is cheaper than meat-on-the-hoof-or-claw, no one will buy the undercut opponent. Slow grown red meat & poultry will vanish from the marketplace, similar to whale oil's flame out when kerosene outshone it in the 1870's. Predictors believe that IVM will sell for half the cost of its murdered rivals. This will grind the $2 trillion global live-meat industry to a halt (500 billion pounds of meat are gobbled annually; this is expected to double by 2050). Bloody sentimentality will keep the slaughterhouses briefly busy as ranchers quick-kill their inventory before it becomes worthless, but soon Wall Street will be awash in unwanted pork bellies.

Special Note: IVM sales will be aided by continued outbreaks of filthy over-crowded farm animal diseases like swine flu, Mad Cow, avian flu, tuberculosis, brucellosis, and other animal-to-human plagues. Public hysteria will demand pre-emptive annihilation of the enormous herds and flocks where deadly pathogens form, after safe IVM protein is available.

Urban Cowboys

Today's gentle drift into urbanization will suddenly accelerate as unemployed livestock workers relocate and retrain for city occupations. Rural real estate values will plummet as vast tracts of ranch land are abandoned and sold for a pittance (70% of arable land in the world is currently used for livestock, 26% of the total land surface, according to the United Nations Food and Agriculture Organization). New use for ex-ranch land? Inexpensive vacation homes; reforested parks; fields of green products like hemp or bamboo. Hot new city job? Techies and designers for In-Vitro Meat factories.

Healthier Humans

In-Vitro Meat will be 100% muscle. It will eliminate the artery-clogging saturated fat that kills us. Instead, heart-healthy Omega-3 (salmon oil) will be added. IVM will also contain no hormones, salmonella, e. coli, campylobacter, mercury, dioxin, or antibiotics that infect primitive meat. I've noted above that IVM will reduce influenza, brucellosis, TB, and Mad Cow Disease. Starvation and kwashiokor (protein deficiency) will be conquered when compact IVM kits are delivered to famine-plagued nations. The globe's water crises will be partially alleviated, due to our inheritance of the 8% of the H2O supply that was previously gulped down by livestock and their food crops. We won't even choke to death because IVM contains no malicious bones or gristle. (Although Hall of Fame slugger Jimmy Foxx choked to death on a chicken bone, about 90% of meat victims are murdered by steak).

Healthier Planet

Today's meat industry is a brutal fart in the face of Gaia. A recent Worldwatch Institute report ("Livestock and Climate Change") accuses the world's 1.5 billion livestock of responsibility for 51% of all human-caused greenhouse gas emissions. Statistics are truly shitty: cattle crap 130 times more volume than a human, creating 64 million tons of sewage in the United States that's often flushed down the Mississippi River to kill fish and coral in the Gulf of Mexico. Pigs are equally putrid. There's a hog farm in Utah that oozes a bigger turd total than the entire city of Los Angeles. Livestock burps and farts are equally odious and ozone destroying. 68% of the ammonia in the world is caused by livestock (creating acid rain), 65% of the nitrous oxide, 37% of the methane, 9% of the CO2, plus 100 other polluting gases. Big meat animals waste valuable land — 80% of Amazon deforestation is done for beef ranching, clearcutting a Belgium-sized patch every year. Water is prodigiously gulped — 15,000 liters of H20 produces just one kilogram of beef. Forty percent of the world's cereals are devoured by livestock. This scenario is clearly unsustainable and In-Vitro Meat is the sensible alternative. (Although skeptics warn that IVM factories will produce their own emissions, research indicates that pollution will be reduced by at least 80%.) Once we get over the fact that IVM is oddly disembodied, we'll be thankful that it doesn't shit, burp, fart, eat, over graze, drink, bleed, or scream in pain.

Economic Upheaval

The switch to In-Vitro Meat will pummel the finances of nations that survive on live animal industries. Many of the world leaders in massacred meat (USA, China, Brazil) have diversified incomes, but Argentina will bellow when its delicious beef is defeated. New Zealand will bleat when its lamb sales are shorn. And ocean-harvesting Vietnam and Iceland will have to fish for new vocations. Industries peripherally dependent on meat sales, like leather, dairy and wool, will also be slaughtered. Hide and leather-exporting nations like Pakistan and Kenya will be whipped, but South Korea will profit on its sales of "Koskin" and other synthetic leathers. Huge plantations of livestock crops (soybeans & corn) in Brazil, USA, Argentina, and China can be replaced with wool substitutes like sisal. Smaller nations that excel in food processing will thrive because they'll export IVM instead of importing tonnage of frozen meat. Look for economic upticks in The Netherlands, Belgium, Denmark, France, and especially Japan, who's currently one of the globe's largest importers of beef.

Exotic & Kinky Cuisine

In-Vitro Meat will be fashioned from any creature, not just domestics that were affordable to farm. Yes, *any animal,* even rare beasts like the snow leopard or Komodo Dragon. We will want to taste them all. Some researchers believe we will also be able to create IVM using the DNA of extinct beasts — obviously, "DinoBurgers" will be served at every six-year-old boy's birthday party.

Humans are animals, so every hipster will try cannibalism. Perhaps we'll just eat people we don't like, as author Iain M. Banks predicted in his short story, "The State of the Art" with diners feasting on "Stewed Idi Amin." But I imagine passionate lovers literally eating each other, growing sausages from their co-mingled tissues overnight in tabletop appliances similar to bread-making machines. And of course, masturbatory gourmands will simply gobble their own meat.

Farm Scrapers

The convenience of buying In-Vitro Meat fresh from the neighborhood factory will inspire urbanites to demand local vegetables and fruits. This will be accomplished with "vertical farming" – building gigantic urban multilevel greenhouses that utilize hydroponics and interior growlights to create bug-free, dirt-free, quick-growing super veggies and fruit (from dwarf trees), delicious side dishes with IVM. No longer will old food arrive via long polluting transports from the hinterlands. Every metro dweller will purchase fresh meat and crispy plants within walking distance. The success of FarmScrapers will cripple rural agriculture and enhance urbanization.

We Stop the Shame

In-Vitro Meat will squelch the subliminal guilt that sensitive people feel when they sit down for a carnivorous meal. Forty billion animals are killed per year in the United States alone; one million chickens per hour. I list this last even though it's the top priority for vegetarians, because they represent only 1-2% of the population, but still... IVM is a huge step forward in "Abolitionism" – the elimination of suffering in all sentient creatures. Peter Singer, founding father of Animal Liberation, supports IVM. So does every European veggie group I contacted: VEBU (Vegetarian Federation of Germany), EVA (Ethical Vegetarian Alternative of Belgium), and the Dutch Vegetarian Society. And PETA, mentioned earlier, offers $1 million to anyone who can market a competitive IVM product by 2012.

My final prediction is this: In-Vitro Meat relishes success first in Europe, partly because its "greener," but mostly they already eat "yucky" delicacies like snails, smoked eel, blood pudding, pig's head cheese, and haggis (sheep's stomach stuffed with oatmeal). In the USA, IVM will initially invade the market in Spam cans and Hot Dogs, shapes that salivating shoppers are sold on as mysterious & artificial, but edible & absolutely American.

One Atom at a Time: Nano

Singularity: Nanotech or AI?

http://hplusmagazine.com/articles/nano/singular
ity-nanotech-or-ai

Josh Hall

The question of the relative roles of nanotechnology and AI in forging the shape of the future has been argued in techno-futurist circles for decades. Eric Drexler mentioned AI as a potentially disruptive technology in his seminal 1986 book *Engines of Creation*, and it was discussed at the very first Foresight conference 20 years ago.

It is generally assumed that a self-improving super-human level of AI is part and parcel of the Singularity, and indeed, such was the basis of I. J. Good's and Vernor Vinge's conception of the "intelligence explosion." But let's assume, for the sake of a scenario, that creating self-improving AI is just a lot harder than we think, and that we aren't going to invent it until well after we have flat-out molecular nanotech with the ability to build fast self-replicating diamondoid nanomachines. What then?

One thing Drexler predicted in "Engines" was that without needing to create true human-level intelligence, automated design systems — narrow as opposed to general AI — would enable the creation of highly complex nanosystems, well beyond the capabilities of mere human designers. How did that prediction pan out? I would have to say that it was so accurate, and happened so soon, that it's taken for granted today. Human designers with only pencil and paper would have no chance of designing, say, a modern computer, or indeed any of today's complex engineered systems. Like many areas, design automation is an area that was once considered AI, but isn't any more.

171

What does a Singularity look like with just nanotech and narrow AI? Let's consider the standard list of transhumanist concerns:

LIFE EXTENSION: Playing around with the interiors of our cells and so forth is clearly a nanotech application. Uploading or radical body improvement is the same.

AI: Ultimately, we get AI by uploading, doing lots of neuroscience, and understanding how the brain works. We get human level AI but not super-intelligent ones. We do ultimately get faster ones, but our uploads can be faster too.

PERSONAL NANOFACTORIES AND UBIQUITOUS WEALTH: Nanofactories wouldn't be quite as powerful without a superintelligence to drive them. They could only make what someone invented and designed, rather than inventing things themselves. But that would be enough to kick the entire physical economy over into a Moore's Law-like growth mode, eradicating hunger and poverty in a decade or two.

FLYING CARS, SPACE TRAVEL, OCEAN AND SPACE COLONIZATION: Again, these are clearly nanotech applications. The modifications to the standard human body necessary to thrive in space require significant nanotech capabilities.

ROBOTS: Robots with human mental capabilities and virtually any physical capabilities would be straightforward, and would rapidly become affordable for everyone.

All of these areas require more scientific knowledge than we have now, but not more than the current rate of scientific progress human scientists are likely to produce in the next few decades. The current techniques of narrow AI are capable of automating pretty much any well defined task, albeit with more programming effort than would be necessary if the machine could learn for itself.

The modifications to the standard human body necessary to thrive in space require significant nanotech capabilities.

With its Moore's Law rates of increasing capability and reducing costs for really high tech physical equipment, one of the things that a nanotech revolution could do is to make scientific instrumentation ever more available. Existing efforts toward open-source science would be enhanced and given more headroom. The scientific knowledge, ingenuity, and experience necessary for the full utilization of the physical capabilities of nanotech could grow as rapidly as the Internet and cell phone use has over the past couple of decades.

So, back to the present. I don't really expect AI to lag behind nanotech as much as this analysis suggests. In fact, I think it will precede it. But even if AI were to stall at roughly its current level of capability, something like a Singularity, a reprise of the Industrial Revolution that boosts our civilization from terrestrial to solar, making us all long-lived, healthy, wealthy, and maybe a little bit wiser, is not only possible but very likely.

How Close Are We
to Real Nanotechnology?

Drexler, Hall, Merkle, Freitas
and the Nanotechnology Roadmap

http://hplusmagazine.com/articles/nano/how-
close-are-we-real-nanotechnology

Surfdaddy Orca

Nanotechnology. You're probably familiar with its promise in the form of the *Star Trek* replicator. You want a glass of white wine and Chicken Cordon Bleu? Punch the keys on the microwave-like replicator and seconds later you'll be drinking a glass of Pinot Gris and digging into a plate of steaming hot chicken. The *Star Trek* replicator creates molecules from subatomic particles and arranges those molecules to form a requested object. Hungry? The replicator first forms atoms of carbon, hydrogen, nitrogen –- arranges them into amino acids, proteins, and cells –- and puts them all together as your wine and chicken.

This is the 22nd Century *Star Trek* vision of nanotechnology – creating molecules and objects from quarks, photons, and neutrinos.

The near-future 21st Century vision of nanotechnology –- detailed in Dr. K. Eric Drexler's seminal book *Engines of Creation* and elaborated in a nanotechnology roadmap — has us assembling molecules using tiny, nanoscale robots with on board molecular computers.

"If we rearrange the atoms in coal we can make diamond," explains nanopioneer Dr. Ralph Merkle. "If we rearrange the atoms in sand (and add a few other trace elements) we can make computer chips. If we rearrange the atoms in dirt, water and air we can make potatoes."

Molecular Manufacturing

Such is the vision of nanotechnology — like the Mr. Fusion unit on the DeLorean time machine in the movie series *Back to the Future*, you can imagine dropping a banana peel into your Mr. Nano unit instead. Rather than the 1.5 GW of power required to power Dr. Emmett Brown's time machine, you get Chicken Cordon Bleu (or a bag of potato chips if you prefer).

A promising method of molecular manufacturing first theorized by Drexler and later tested by Merkle and collaborator Rob Freitas is diamondoid mechanosynthesis. This is the mechanical synthesis of stable chemical structures using covalently bonded carbon –- otherwise known as diamond.

This manufacturing technology would build macroscale products atom-by-atom using bottom-up assembly, in contrast to the top-down assembly that represents almost all present-day manufacturing.

Or imagine targeted anti-aging therapy and drug delivery. Nanoscale machines in the bloodstream can be used to target and repair cancer cells and other pathologies. Rob Freitas has written extensively on the medical applications of molecular nanotechnology and medical nanorobotics.

Another promising molecular-level manufacturing technology that can also be used for targeted drug delivery is found in Paul Rothemund's work folding stringy DNA molecules into tiny, two-dimensional patterns known as DNA origami. [See "Swan in a Box –
http://hplusmagazine.com/articles/nano/swan-box]

Like diamondoid mechanosynthesis, DNA origami is a bottom-up fabrication technique that exploits the intrinsic properties of atoms and molecules to make simple nanostructures.

The pace of development toward these nanotech visions is accelerating. Here's the mind-boggling list of recent nanotech developments.

- DNA wrapped carbon nanotubes for artificial tissue

- Two ways to make large scale three dimensional structures out of DNA

  1. Routing a single-stranded scaffold DNA (a virus genome) through every section of a tube template

  2. DNA used to assemble sheets of metal nanoparticles, that could be the basis of nanocircuits and could be integrated with the 3D nanotechnology

- Large scale 3D nanotechnology with DNA that is integrated with carbon nanotubes, diamond nanorods, nanoparticle metal, graphene and other DNA compatible chemistry

- DNA to detect pathogens and be used for drug delivery

- Modified RNA so that it reliably enter cells for drug delivery

- Ultrathin diamond nanorods – only twice as thick as the diamond rod logic for the molecular computers described by Drexler

Given the level of extremely rapid discovery and experimentation, it seems like we may be reaching a tipping point – and it certainly supports the case for Kurzweil's Law of Accelerating Returns.

Foresight Institute

The Foresight Nanotechnology Institute is the leading think tank and public interest institute for nanotechnology. Founded in 1986, this institute was the first organization to provide outreach concerning the benefits and risks of nanotechnology.

The concept of nanotechnology is largely credited to a talk given by physicist Dr. Richard Feynman, a Nobel Prize winner in Physics, at an American Physical Society meeting at Caltech on

December 29, 1959. Feynman described a process to manipulate individual atoms and molecules.

The Foresight Institute established the Feynman Grand Prize in 1996 to motivate scientists and engineers to design and construct a functioning nanoscale robotic arm with specific performance characteristics. This $250,000 incentive prize is somewhat analogous to the Loebner Prize for Artificial Intelligence (AI) – a grail of sorts. Either a human-level AI agent or a nanoscale robot arm will bring big changes.

The Foresight Institute is also helping to set the agenda for the beneficial applications of nanotechnology:

- Providing Renewable Clean Energy
- Supplying Clean Water Globally
- Improving Health and Longevity
- Healing and Preserving the Environment
- Making Information Technology Available To All
- Enabling Space Developmen

Nanotechnology Roadmap

So how do we get there from here? Supported through grants to the Foresight Institute by the Waitt Family Foundation (founding sponsor) and Sun Microsystems, and with direct support from Nanorex, Zyvex Labs, and Synchrona, a distinguished multi-disciplinary group of scientists and engineers have created a 198 pp. nanotechnology roadmap.

The roadmap's executive summary describes a unique, cross-disciplinary process for exploring current capabilities and near-term opportunities in atomically precise technologies (APT), and examines pathways leading toward advanced atomically precise manufacturing (APM).

The long-term vision is the fabrication of a wider range of materials and products with atomic precision. However, there are strong differences of opinion on how rapidly this will occur. All agree, however, that expanding the scope of atomic precision will dramatically improve high-performance technologies of all kinds, from medicine, sensors, and displays to materials and solar power.

If nanotechnology follows Moore's law (transistors on a chip double every 18 months), this level of nanotechnology could occur in the next 15 years or less. The vision includes:

- Precisely targeted agents for cancer therapy
- Efficient solar photovoltaic cells
- Efficient, high-power-density fuel cells
- Single molecule and single electron sensors
- Biomedical sensors (in vitro and in vivo)
- High-density computer memory
- Molecular-scale computer circuits
- Selectively permeable membranes
- Highly selective catalysts
- Display and lighting systems
- Responsive ("smart") materials
- Ultra-high-performance materials
- Nanosystems for APM

The Roadmap identifies three successive development horizons – with timelines clearly dependent upon funding and R&D (Table 1).

Table 1. Projected nanotechnology applications for successive development horizons

SEE ACTUAL ARTICLE AT:

http://hplusmagazine.com/articles/nano/how-close-are-we-real-nanotechnology

So How Close Are We Really?

Given such rapid recent technological progress, how close are we to achieving the vision of molecular nanotechnology in the roadmap? *H+* Magazine asked this question of some of the pioneers who founded the science of nanotechnology: Drexler, Hall, Merkle, and Freitas

K. Eric Drexler

K. Eric Drexler, Ph.D., is a researcher and author whose work focuses on advanced nanotechnologies and directions for current research. Drexler was awarded a Ph.D. from the Massachusetts Institute of Technology (MIT) in Molecular Nanotechnology (the first degree of its kind). Dr. Drexler serves as Chief Technical Advisor to Nanorex, a company developing open source design software for structural DNA nanotechnologies.

To describe progress toward advanced nanotechnology, I'd like to offer an analogy to spaceflight. For decades, there was progress in rocketry, but no spaceflight. Rockets flew faster, higher, and further, yet there were no communications satellites, weather satellites, or astronauts. Within five years of the first satellite launch, there were all three.

Today, there is rapid progress toward nanotechnologies at the level I described in *Engines of Creation* (with updated objectives, of course), but there is a threshold ahead, like reaching orbit, that hasn't yet been achieved. Although the world-changing technologies accessible at that level are still years in the future, there is a technology roadmap, and progress is accelerating.

J. Storrs Hall

"Josh" Hall is President of Foresight Institute, author of Nanofuture: What's Next for Nanotechnology, fellow of the Molecular Engineering Research Institute and Research Fellow of the Institute for Molecular Manufacturing. Hall was also a computer systems architect at the Laboratory for Computer

Science Research at Rutgers University from 1985 until 1997. His most recent book is Beyond AI: Creating the Conscience of the Machine.

There is certainly a lot of progress in the labs. The DNA origami (and 3D structures) work is quite promising. Give them a few years and they will be making circuits by having the DNA mats bind and position components and wires. I'd guess that by 2020 the self-assembly and top-down photolithography approaches to building circuits will have merged. The semiconductor industry roadmap calls for four nanometer gate lengths in 2022, which is just about right...

The big roadblock that nobody's looking at is powered moving parts – nanomachines. It's a toss-up as to when that will be hurdled, because it depends so much on how much effort is put into it. The Foresight Institute is going to be concentrating on this in the coming years... to what effect we don't know. My best guess is that the technology to win the Feynman Grand Prize (that is, build a working robot arm at the nanoscale) will come somewhere between 2020 – 2030, depending on the level of effort.

Ralph Merkle

Dr. Merkle received his Ph.D. from Stanford University in 1979 where he co-invented public key cryptography. He joined Xerox PARC in 1988, where he pursued research in security and computational nanotechnology until 1999. He was a nanotechnologytTheorist at Zyvex until 2003, when he joined the Georgia Institute of Technology as a Professor of Computing until 2006. He is now a Senior Research Fellow at the Institute for Molecular Manufacturing.

There's a lot of work on self assembly in general and DNA nanotechnology in particular – it's very exciting and advances the state of the art. We can do things today at the molecular scale that we couldn't do even a few years ago.

I've been working with (Rob) Freitas on mechanosynthesis, which is one of several routes to develop nanotechnology that is being explored. We've been collaborating with various people, including Professor Philip Moriarty of the Nanoscience Group in the School of Physics at the University of Nottingham, an experimentalist in England.

It might still be a few decades before we have the ability to inexpensively arrange atoms in most of the ways permitted by physical law, but progress has been remarkable — the exponential trends continue to be exponential, and there's every reason to think they will continue that way for some time.

Rob Freitas

Robert A. Freitas Jr. is Senior Research Fellow at the Institute for Molecular Manufacturing (IMM) in Palo Alto, California, and was a Research Scientist at Zyvex Corp., the first molecular nanotechnology company, during 2000–2004. Freitas is the author of Nanomedicine, the first book-length technical discussion of the potential medical applications of molecular nanotechnology and medical nanorobotics.

I think the pace of nanotechnology development is slowly accelerating on all fronts.

Progress in DNA nanotech continues to quicken, increasing our ability to arrange organic matter by design at the molecular level.

Diamondoid MNT (Molecular Nanotechnology) also continues to make steady progress, both in terms of experimental and theoretical accomplishments, with the first focused experimental work on diamond mechanosynthesis finally underway since last October.

We are clearly moving closer to achieving the long-term vision of medical nanorobotics and MNT.

In Conclusion: The Coming Diamond Age

Surprisingly, a work of SF is a required textbook at MIT's Media Lab.

The Diamond Age by Neal Stephenson highlights the world of the relatively near future. "The Feed" allows most anything to be created at any outlet (think Star Trek replicators), creating a minimum standard of living for all mankind through the spread of nanotechnology.

Nanoscale clouds of mites (think Utility Fog) – engineered nanoprobes – fly about gathering information. The mites can examine people from the inside and detect medical problems. Mites, like viruses, can infect or inoculate people.

Stephenson's *The Diamond Age* is, of course, based on diamondoid mechanosynthesis – the ability to assemble diamond structures form carbon atoms that Merkle and Freitas are actively researching today.

It's notable that both Drexler and Merkle are given honorable mention in this SF-textbook-blueprint for the 21st Century.

If we are to believe Drexler, Hall, Merkle, and Freitas and the nanotechnology roadmap – and the accelerating pace of nano-technology development does appear to support them – we are on the cusp of a new era in the next 20-30 years.

Such an era would be akin in magnitude to the Information Age, the Industrial Age, the Iron Age, the Bronze Age, and the Stone Age.

Except this time, it's the Diamond Age.

Targeting Cancer Cells
with Nanoparticles

http://hplusmagazine.com/articles/nano/targetin
g-cancer-cells-nanoparticles

Surfdaddy Orca

Cancer is the number two cause of death in the United States, killing over 550,000 people annually according to the American Cancer Society's 2009 *Cancer Statistics* report. Although there has been significant progress in the treatment of cancer over the past 20 years, the overall five-year survival rate is still only 66%. Some of the causes for this high mortality rate include tumor identification and classification, late diagnosis, and incomplete or inadequate treatment – all of which lead to poor patient outcomes.

The majority of promising new cancer therapies target selected cancer-activated biochemical pathways. To maximize the clinical utility of these new therapies, researchers must identify and understand the molecular characteristics of each patient's tumor, and to obtain as much diagnostic and prognostic information as possible for each patient.

Enter nanomedicine. Led by Elena Rozhkova, scientists from the U.S. Department of Energy's (DOE) Argonne National Laboratory and the University of Chicago's Brain Tumor Center have developed the first nanoparticles that seek out and destroy glioblastoma multiforme (GBM) brain cancer cells without damaging nearby healthy cells.

Nanomedicine, an offshoot of nanotechnology, refers to highly specific medical intervention at the molecular scale for curing disease or repairing damaged tissues, such as bone, muscle, nerve, or brain cells. Nanoparticles — anywhere from 100 to 2500 nanometers in size — are at the same scale as the biological molecules and structures inside living cells. Cancer detection using nanoparticles shows great promise as a therapy for certain types of cancer. And the U.S. National Institute of Health (NIH) is taking nanoparticles very seriously. The NIH has established a national network of eight Nanomedicine Development Centers, which serve as the intellectual and technological core of the NIH Nanomedicine Roadmap Initiative.

Dr. Rozhkova's solution involves chemically linking titanium dioxide nanoparticles to an antibody that recognizes and attaches to GMB cells, reports *Science Daily*. When they exposed cultured human GMB cells to these so-called "nanobio hybrids," the nanoparticles killed up to 80 percent of the brain cancer cells after 5 minutes of exposure to focused white light. The results suggest that these nanoparticles could become a promising part of brain cancer therapy, when used during surgery.

GMB is a particularly nasty form of cancer that often causes death within months of diagnosis. Titanium dioxide nanoparticles, a type of light-sensitive material widely used in sunscreens, cosmetics, and even wastewater treatment, can destroy some cancer cells when the chemical is exposed to ultraviolet light even without the use of antibodies. However – until now – researchers have had difficulty getting nanoparticles to target and enter cancer cells while avoiding healthy cells.

Titanium dioxide is not the only nanoparticle that shows promise in cancer therapy. Gold nanospheres – nearly perfectly spherical nanoparticles that range in size from 30 to 50 nanometers – are being used to search out and "cook" cancer cells. The cancer-destroying nanospheres show promise as a minimally invasive future treatment for malignant melanoma, the most serious form of skin cancer. Melanoma now causes more than 8,000 deaths annually in the United States alone and is on the increase globally.

The hollow gold nanospheres are equipped with a special peptide that draws the nanospheres directly to melanoma cells, while avoiding healthy skin cells. After collecting inside the cancer, the nanospheres heat up when exposed to near-infrared light, which penetrates deeply through the surface of the skin, explains study co-author Jin Zhang, Ph.D., a professor of chemistry and biochemistry at the University of California in Santa Cruz. This procedure is a variation of photothermal ablation, also known as photoablation therapy, a technique in which doctors use light to burn tumors. Since the technique can

destroy healthy skin cells, doctors must carefully control the duration and intensity of treatment.

"This technique is very promising and exciting," says Jin Zhang. "It's basically like putting a cancer cell in hot water and boiling it to death. The more heat the metal nanospheres generate, the better."

Photoablation therapy can be greatly enhanced by applying a light absorbing material such as metal nanoparticles to the tumor. However, many materials show poor penetration into cancer cells and limited heat carrying capacities. These materials include solid gold — as opposed to the hollow nanospheres used by Dr. Zhang and his colleagues. Other materials such as solid nanorods lack the desired combination of spherical shape and strong near-infrared light absorption required for effective photoablation therapy.

Dr. Zhang and his colleagues worked with the hollow gold nanospheres — each about 1/50,000th the width of a single human hair — to develop more effective cancer-burning materials. Previous studies by others suggest that hollow gold nanospheres (nanoshells) have the potential for strong near-infrared light absorption. However, until very recently, researchers have been largely unable to produce them successfully in the lab.

The following video shows how light-absorbing gold nanoparticles can also use epidermal growth factor receptors (EGFR) sites for cancer cell detection with simple white light:

See:
http://www.youtube.com/watch?v=uyhxRIvw_cY

NIH expresses the hope that nanomedicine — including the use of nanoparticles and eventually nanobots (see the *H+* article "Nanobots in the Bloodstream" in Resources) — will allow scientists to build new devices for a wide range of biomedical applications. According to the NIH Roadmap, such technology would be used for "detecting infectious agents or metabolic imbalances with novel, tiny sensors, replacing 'broken' machinery inside cells with new nanoscale structures, or generating miniature devices that search for, and destroy, infectious agents."

A second phase for the NIH nanomedicine program was recently approved. "Centers will continue to expand knowledge of the basic science of nanostructures in living cells, will gain the capability to engineer biological nanostructures, and then will apply the knowledge, tools, and devices to focus on specific target diseases," reports the NIH Roadmap.

The idea is to apply "acquired fundamental knowledge" such as the experimental results obtained using titanium dioxide nanoparticles and hollow gold nanospheres to actually treat disease. Such targeted cancer therapies may ultimately ease the despair of millions of cancer victims worldwide.

Engineering an End to Aging

http://hplusmagazine.com/articles/forever-young/engineering-end-aging

Michael Anissimov

Age-defying creams and lotions, esoteric herbs and elixirs, Botox and plastic surgery -- what do they all have in common?

None of them will actually increase your life span. Usually, they're snake oil. At best, they improve external appearance without actually extending life. We deserve better, and we'll need it if we want to live longer than the typical three score and ten years.

The first thing to realize is that nature doesn't specifically want us to die. There is no "death gene." For any species in any environmental context, there is an ideal lifespan from an adaptive point of view — an evolutionary optima. One evolutionary strategy includes species that reproduce quickly and die off fast. Another includes species that reproduce slowly and live for a long time. Call it quality versus quantity. Thankfully for humans, we're squarely in the quality column, but many would agree that 80 to 90 years is not enough.

We perish not because of some internal clock that says, "Time to die now!," but because of a lack of attention and self-healing... mere neglect. Once we've reproduced a few times, in the eyes of nature, our usefulness has run its course. We are cast aside, onto a pile of skeletons 600 million years deep. This is unacceptable, and we need to find a new way, but since nature isn't actively working against us — just neglecting us — the challenge is surmountable.

Longevity in Nature

First, let's look to nature for inspiration. Are there any animals with extraordinarily long life or regenerative capacities? Absolutely.

There is one animal that scientists believe is immortal — the lowly hydra, a simple, microscopic freshwater animal shaped something like a tiny squid. Apparently the challenges of indefinite tissue regeneration are simple enough for such a small organism that nature has solved them. American biologist Daniel M. Marinez did a study of mortality in three colonies of hydra for four years straight, and barely any of them died.

Death rates were random, uncorrelated with age. This means they weren't displaying senescence (aging), and died from other causes. In almost all other known species, death rates increase with age. Not in hydra. They die from getting eaten, or infected by a virus, or squished, but not from aging. There could be a thousand-year-old hydra out there, maybe in a small lake right in your neighborhood. We don't know, because there is no way of telling their age by looking at them!

Planarians – those odd animals that look like a slug squished in a microscope slide – are another organism that scientists suspect may be immortal. No detailed studies have been conducted yet. In many cases, if you cut a planarian in half, it becomes two planarians. These live as long as one born by conventional means. If you kept cutting a planarian in half, it might never die, because each piece would go on living.

What about more complex animals? There are our friends in the order Testudines: turtles, tortoises, and terrapins. Scientists have examined the internal organs of young and old turtles and found that they look exactly the same. Something in a turtle's physiology prevents these organs from breaking down. An article in *Discover* magazine asked, "Can Turtles Live Forever?" and came to the conclusion that it's entirely possible. Like hydra, turtles experience no increase in mortality rates and no decrease in reproductive rates as they grow older. There are turtles 150 years old that exhibit no signs of aging. Harriet the Turtle, a pet of Charles Darwin's, was born in 1830 and died only in 2006. It seems turtles can die from disease, injury, or predation, but not aging. This quality is called "negligible senescence." Sign me up.

From these animal examples, we see it would be premature to state that negligible senescence is biologically impossible, as is frequently assumed. Nature seems to be uninterested in our quaint notion that all organisms must age. The question is… how can we make this work for humans? The oldest person who ever lived, Jeanne Louise Calment, kicked the bucket at the age of 122 1/2. Can we push that boundary?

Engineering Negligible Senescence

Enter Dr. Aubrey de Grey, a biogerontologist from the UK, and his "strategies for engineered negligible senescence" (SENS) plan. Instead of exclusively studying the complex biochemical processes of aging in detail, as in gerontology, or ameliorating the worst symptoms of age-related decline, as in geriatrics, de Grey and his supporters advocate an "engineering approach" to aging that asks, what are the main categories of age-related biochemical damage, and how can we fix them? The idea is not to eliminate the sources of age-related damage, but to fix the damage fast enough so it doesn't accumulate and cause health problems. This is far easier than deciphering all the intricacies of the biochemistry of aging.

Although some tentative engineering approaches to aging had been proposed before, it was de Grey who really fleshed it out, popularized it, and made it respectable. It's no wonder that he has already raised $10 million in funding for his organization, the Methuselah Foundation.

As de Grey points out, gerontologists have discovered seven biochemical causes of aging. The last cause was discovered in 1981, and considering how immensely far our knowledge of biology has come since that time, it seems quite likely that these seven causes are all of them. De Grey calls these causes of aging the "Seven Deadly Things." They are: (1) cell loss, (2) death resistant cells, (3) nuclear DNA mutations, (4) mitochondrial DNA mutations, (5) intracellular junk, (6) extracellular junk, and (7) extracellular crosslinks. That's it. If we find medicines or therapies that can clean up this damage, we could extend our lifes pans to great lengths and achieve negligible senescence in humans.

A word on a philosophical point of view: many world philosophies and religions teach, or strongly imply, that the body depends on some immaterial animating force, a soul or chi, to give it life. Scientists disagree: the functioning of the body seems entirely rooted in atoms, molecules, and forces between them. As recently as 1907, French philosopher Henri Bergeson wrote about an élan vital, or vital force, that animated all living things and drove their evolution and development. This was closely connected to the idea, common at the time, that organic molecules could not be synthesized by inorganic precursors. Unfortunately for Bergeson and other vitalists, Friedrich Wöhler, the father of biochemistry, had already synthesized urea from inorganic precursors as early as 1828, and scientists were becoming more and more convinced that the same laws of biochemistry that govern inorganic molecules govern organic molecules as well.

Because the laws of chemistry apply to both life and non-life, aging is an entirely chemical, non-mystical process of degradation with specific physical causes. Although it is a matter of preference whether you consider aging a "disease" or not, from the perspective of the body, aging is like a disease — a life-destroying biochemical phenomenon occurring in the body. And like diseases, aging is treatable.

It is due to the complexity and the aura of inevitability around aging that people have only recently begun to look at it this way. Some say that aging is something mandated by God, and we have no right to mess with it, but these very same people have used this same argument throughout history to protest against vaccinations, the dissection of cadavers, organ transplants, and numerous other therapies or techniques of extreme medical value. Is it so radical to say that being healthy is a good thing, and that we should use whatever ethical strategies are available to pursue that end?

Aubrey de Grey's SENS plan is complex and quite thorough. To examine it in full, I suggest looking at the website of the Methuselah Foundation, or getting his recent book, *Ending Aging*. But I will summarize the basics here.

The first cause of aging is cell loss, or cell atrophy. For most of our lives, our bodies are programmed to replace cells when they die. Our individual cells live much shorter life spans than the body itself: some cells last a few years, others, like skin cells, a few weeks. All of them are constantly regenerated using the body's supply of stem cells. Over time, the processes of cell replenishment begin to break down. This is what causes muscle atrophy among the old, and the phenomenon especially afflicts the heart and brain, our two most important organs. To fix this problem, two strategies have been proposed: stimulating the division of existing cells, or introducing new cells, possibly including stem cells. Both are under investigation.

The second cause of aging is death resistant cells, cells that overstay their welcome. There are three main types of cells guilty of this offense. The first are visceral fat cells, fat cells that build up around our internal organs. These cause a progressive loss in our body's ability to respond to nutrients from the stomach. Eventually, it leads to Type 2 Diabetes. The second type of cells is called senescent cells, cells that have lost the ability to reproduce. These stick around, releasing proteins that are dangerous to their neighbors. Thankfully, they primarily aggregate in just one type of tissue, the cartilage between our joints. A third type is a category of immune cells called "memory cytotoxic T cells." These build up faster than other immune cells and refuse to go away, crowding out the other immune cells and eventually causing disease. There are two approaches to solving these problems: inject something that makes the unwanted cells commit suicide but doesn't touch other cells, or stimulate the immune system to kill the target cells.

The third cause of aging is mutations in the DNA of the nucleus, the center of every cell. Most of these mutations are entirely harmless, as they only affect a few cells at a time. These cells eventually die and are replaced with unmutated cells. Mutations get dangerous when they lead to malignant cells that self-replicate — otherwise known as cancer. So, finding a cure for a cancer is a subtask of finding a cure for aging. According to de Grey, this is the most difficult part of the strategy, because cancer is constantly evolving to exploit us.

There are several proposed approaches to finding a cure for cancer, but de Grey's favored strategy is one called "Whole-body Interdiction of Lengthening of Telomeres" (WILT). The Methuselah Foundation's website calls WILT "a very ambitious but potentially far more comprehensive and long-term approach to combating cancer than anything currently available or in development." It is based on a vulnerability shared among all cancer cells: their need to renew their telomeres, junk DNA that serves as the ends of chromosomes. Telomeres of a certain length are necessary for a cell to self-replicate. If the telomeres are too short, the cell self-destructs. When cancer hijacks the body's cells, the cancer cells replicate so rapidly that their telomeres shorten quickly. The cancer cells avoid destruction by using the cell's protein synthesis machinery to build enzymes -- telomerase and ALT — that extend telomeres, and allow endless selfreplication. Previous attempts at cancer cures target these enzymes, but WILT proposes removing the very genes that contain the information necessary to synthesize them.

Removing the genes underlying the synthesis of telomerase will mean that all cancers will self-destruct before becoming a serious problem to their host, effectively curing cancer. This is one of the most ambitious strands of the SENS plan. The challenge of this approach is that removing these genes in all the tissues of the body will mean that the body's natural cells will have a limited lifespan, as they will not be capable of lengthening their telomeres. To counteract this will require introducing stem cells with renewed telomeres into the body every decade or so. This has already been demonstrated in mice with cells of the blood and gut. Skin and lungs will be next. When this therapy is used to cure cancer in mice, tremendous resources will be pumped into efforts to develop a therapy that works for humans.

The fourth cause of aging is mutations in the mitochondria, the "power stations" of the cell. Mitochondria have their own DNA, much less than that which is in the nucleus of the cell, but some of it is essential to synthesizing the proteins that make it up. When the DNA is damaged, the mitochondria break down. Mitochondrial DNA is especially susceptible to damage because of two reasons. The first is that mitochondria, being the site of cellular respiration, are heavily exposed to its by-products — dangerous free radicals. These react with the DNA, causing it to mutate. The second is that mitochondria lack the complex DNA-repair machinery found in the nucleus.

Luckily, although mitochondria are made of thousands of proteins, only 13 of them are synthesized using the genes of the mitochondria itself. The rest are synthesized in the nucleus and imported in. The solution to this problem is to move the thirteen critical genes from the mitochondria to the nucleus of the cell. Evolution has already been doing this without our help for millions of years, and we need to finish the job. This will require using gene therapy to add supplementary genes. Gene therapy is in its early stages, but it has been used effectively to replace defective genes with functional ones, helping cure genetic diseases. Research is under way to improve the process and test it with mice.

The fifth cause of aging is intracellular junk. Cells synthesize, reconstruct, and deconstruct many thousands of different molecules during the course of their operation. Every once in a while a cell ends up with a molecule so large or unusual that it has trouble breaking it up. If a molecule can't be broken down by the "incinerator" of the cell, the lysosome, it stays there forever. In cells that don't divide, this can build up to critical levels. This includes some cells in the heart, the back of the eye, some nerve cells, and white blood cells trapped in the walls of arteries. This can cause diseases, such as Alzheimer's, Parkinson's, macular degeneration (the leading cause of acquired blindness), and atherosclerosis. To clean up intracellular junk, the SENS project proposes equipping the lysosome with new enzymes, thereby expanding the range of molecules it can break down, allowing it to digest even very large or unusual molecules.

The sixth cause of aging is extracellular crosslinks, molecular garbage that accumulates outside cells, linking together proteins that otherwise slide smoothly over each other. These can lead to some of the most outwardly visible effects of aging: wrinkles in tissue and the like. Fortunately, these crosslink molecules have chemical structures different than the healthy tissue of the body, so it shouldn't be too hard to find an enzyme that breaks them down while leaving the rest alone. In fact, just one type of crosslinks, called glucosepane crosslinks, may count for up to 98% of all long-lived extracellular crosslinks in the human body, meaning if we figure out a way to get rid of these, we'll have almost solved this cause of age-related damage.

The seventh and last known cause of aging is general extracellular junk, the type that just floats around instead of linking together proteins. Most of these junk molecules are called amyloids, and they build up in everyone, but are especially found in the brains of Alzheimer's patients. The main approach to dealing with this, already being pursued by at least one company, is to stimulate the body's immune cells to clear out these molecules. There is a strong overlap between treatments for Alzheimer's and atherosclerosis and anti-aging treatments that address this cause, so there seems to be significant momentum in the right direction.

There may be other causes of aging that emerge after we have solved most of these seven. We'll just have to wait and see. But if all these seven causes of aging were eliminated, people could live a lot longer — maybe even hundreds of years. That would buy us more time to develop new therapies to address the remaining sources of aging.

It's hard to imagine why we wouldn't want to fight the scourge of aging — besides killing more than 100,000 people per day; it makes us suffer for years or decades before it kills us. Everyone is susceptible. Instead of seeing aging as inevitable, why don't we view it as a disease and search for a cure?

Smart Biology to the Rescue

http://hplusmagazine.com/articles/forever-young/smart-biology-rescue

Alex Lightman

As the Boomers begin to go gray and fragile, those with way high expectations confront an uncomfortable fact — nobody has done much about aging, throughout their lifetimes... And they get angry.

How could this be?! Technology has carried us along on its broad back, giving us computers, conveniences, Internet and media wonders. But aches and pains foretell much bad news ahead. We can do better, but to do it we'll have to *reinvent biology*.

Face it: young or old, we can't solve "the aging problem" using the standard 20th Century research methods of cell biology. Sure, they had some great success with some other medical problems — nobody fears smallpox, polio and other old school diseases. Diet and exercise help, too, But nobody has done much *directly* about the mechanisms that erode our bodies.

Why? Because beyond our twenties, natural selection doesn't help us much. Once we start reproducing, all the genes that break us down get passed on to the next generation. It's been that way throughout natural evolution. Aging arises from a lack of natural selection in later adulthood.

So what's the smart biology dodge around this? Make selection work for us by forcing it to produce longer-lived animals. And then learn from what forced selection tells us. That's what the 21st Century medicine man knows. We can already see him peeking around the corner up ahead. He says: *Your aging*

199

comes from multiple genetic deficiencies, not a single biochemical problem.

Michael Rose at UCI saw all this coming 30 years ago. He started breeding longer and longer lived fruit flies (Drosophila) by having them not keep their eggs to make the next generation until much later in adult life. Do that in humans and you'll be trying to get babies out of sixty years olds — not a promising route — though I guess the Italians, with a 1.1 fertility rate (2.1 is replacement) are trying.

Rose's years of painstaking Methuselah fly stud-servicing produced a fracking miracle: flies that live 4.5 times longer than ordinary flies. Do that with humans for 10,000 years — 500 generations — and you'll start approaching Rose's results. But to get the advantages today you'd have to have started back before there were cities.

That's why smart biology uses "animals" — particularly insects, that don't live long — to squeeze those 10,000 years down to a career lifetime of about 40 years. (Rose is in his 50s.)

What do the genetic inventories of these Methuselah flies show? Multiple, overlapping genomic pathways. About 75% of the genes do the same jobs in flies as they do in humans. We share these basic operating systems with insects that we parted company from about a billion years ago. (Yes, intelligent design fanatics, you are related to mosquitoes. Suck it up. Stop bugging me.)

Genescient Corporation acquired the use of the Methuselah flies' genomics and has developed their implications for three years. Knowing that these complex genomic pathways can enhance resistance to the many disorders of aging, their crucial step is to find substances that can enhance the action of those pathways. Designer supplements containing nutrients made using detailed genomic information — a field called *nutrigenomics* — are about to come to your local supermarket, some of them using obscure traditional medicines. This is the essence of a 21st Century approach to aging. Nothing like it has existed before this year.

Noted hard science fiction author and Genescient (which means 'smart genes') cofounder Gregory Benford argues that there seems to be no fundamental reason why we can't live to 150 years or even longer ("and you can have sex up to 150 also"... I *like* that part). After all, nature has done quite well on her own, using pathways humans share and can now understand. The 4,800-year-old bristlecone pine, and koi fish over 200 years old, attest to this, not to mention tortoises.

billion years developing these pathways. Genescient aims to explore them rather like someone playing *SimEarth* or *Spore*: by speeding up generational times. The medical technology emerging now acts on these basic pathways to immediately affect all types of organs. Traditionally, medicine focuses on disease by isolating and studying organs, and organizes diseases mostly by spotlighting local disorders. Genomics can focus on entire organisms by looking at the entire picture.

You'll know 21st Century medicine has arrived when you see immortality pills featuring mixes of designer supplements.

You'll know 21st Century medicine has arrived when you see immortality pills featuring mixes of designer supplements. These will regulate your own genes to improve their resistance to the many ways things can go wrong. The plausible outcome of taking these pills will be bodies that don't seem to age as fast and that can maintain vigor long after the childbearing years, when we traditionally begin to show wear and tear.

That's what happened with Michael Rose's Methuselah flies. The Genescient labs track fly vigor by their mating frequency. They count how often the flies get it on — and the numbers of eggs the females lay. Those horny Methuselahs beat out the other flies in the mating game. Basically, the more you want and get sex, the longer you will live. Adult Friend Finder and Be Naughty, you are free to quote me on this.

After the first wave of designer supplements, we'll see customized nutrigenomic pills. Medicine will get tailored to each personal genome. Targeting a person's own suite of complex pathways, smart designer supplements and drugs can propel the repair mechanisms and augmentations that nature provided. This will benefit everyone, not just the genomically fortunate.

The 21st Century has scarcely begun, and already it looks as though many who welcomed it in will see it out. The first person to live to 150 may be reading this right now.

Alex Lightman is the author of the first book on 4G wireless, Brave New Unwired World (Wiley) and founder of pioneering companies in 3-D and Hollywood websites, wearables, and IPv6. He welcomes friending on Facebook.

I Am Ironman!: HAL (Hybrid Assistive Limb) Cybernetic Suit

http://hplusmagazine.com/articles/robotics/i-am-ironman

Tristan Gulliford

Cyberdyne Corporation of Japan, in conjunction with Daiwa House, has begun mass production of a cybernetic bodysuit that augments body movement and increases user strength by up to tenfold.

The HAL (Hybrid Assistive Limb) suit works by detecting faint bioelectrical signals using pads placed on specific areas of the body. The pads move the HAL suit accordingly. The Cyberdyne website explains: "When a person attempts to move, nerve signals are sent from the brain to the muscles via motoneuron, moving the musculoskeletal system as a consequence. At this moment, very weak biosignals can be detected on the surface of the skin. HAL catches these signals through a sensor attached on the skin of the wearer. Based on the signals obtained, the power unit is controlled to wearer's daily activities."

Among the potential applications, Cyberdyne is emphasizing helping people with movement disabilities, augmenting strength for difficult industrial tasks, disaster rescue, and entertainment.

The HAL suit is not currently available only in Japan. But according to *Nikkei News,* it should be marketed at approximately $4,200 US dollars when it reaches the market.

Watch the HAL suit in action:
http://www.youtube.com/watch?v=yIInxgyilEA

My New Sense Organ

http://hplusmagazine.com/articles/enhanced/my-new-sense-organ

Quinn Norton

I am beta testing a new sense. My new sensory organ is a small anklet strap with a LiPo battery and circuit board attached to an electronic compass on the anklet's side. Inside the strap are eight small buzzers that are up against my skin. As I sit here typing, the buzzer on the very left side of my left ankle is gently informing me which way is north. The anklet is called a Northpaw. My new sense is perfect direction.

The Northpaw is based on the Feelspace, a project organized by the Cognitive Psychology department of Universität Osnabrück in Germany. The principle is simple and elegant. The buzzers signal north to the wearer. The wearer gets used to it, often forgetting it's there. They just start getting a better idea of where they are through a kind of subconscious dead reckoning. It started as a university experiment. They got the data, wound it up, and never intended to commercialize it.

Adam Skory liked the idea so much he wanted to make one for himself. He teamed up with some friends at the San Francisco hackspace Noisebridge and built it. In the process they decided that they might as well sell kits so others could make it more easily. Skory gave me version 1, and set me loose in San Francisco.

I think of myself as having a good sense of direction, and I do, in a way. It's just wrong most of the time. My north drifts quite far from magnetic north. But it's a consistent wander, still useful for navigation, if patently untrue. The Northpaw isn't perfect either. This early version has the occasional bug and misplaced buzz, but it's better than I turn out to be. I had wrong assumption I didn't know about, my confidence in my cognitions misplaced.

It doesn't work while driving because the compass doesn't like being turned on its side, as it is when you work car pedals. Magnetic fields mess it up (of course) and I can feel it circling my foot on escalators or seeming to vacillate directions randomly as I rest my foot on the floor of the subway. But that's interesting too — to feel the specific places where infrastructure interferes with the Earth's magnetic field.

I returned home to Washington DC to find that, far worse than my old haunt San Francisco, my mental map of DC swapped north for west. I started getting more lost than ever as the two spatial concepts of DC did battle in my head. Eventually, the Northpaw won, and the NW/NE/SW/SE on DC street signs started making a whole lot more sense.

My relationship with the Northpaw is still shaky. It passes in and out of my integrated experience. When it's at its best, my awareness is not of the touch from the Northpaw, it's the awareness of north from the Northpaw. I make it dance around by spinning my office chair, but it doesn't keep up. I get nauseous and dizzy much quicker wearing the Northpaw than I do spinning without it.

The Northpaw experience has been more about realigning my reality than about its being useful. It tells me more about the world, rather than giving me immediately practical information. But then, I have more a Google Maps than compass lifestyle.

Skory told me that in the time he was wearing his Northpaw he found that hiking trails were much more twisted that he thought they were. But even straight things aren't that straight. I find roads and paths drifting in ways I never noticed. Not always... Not a lot, but just enough to be unsettling. My world's Euclidian consistency is becoming questionable.

Sports Enhancement and Life Enhancement: Different Rules Apply

http://hplusmagazine.com/articles/enhanced/spor
ts-enhancement-and-life-enhancement-different-
rules-apply

Quinn Norton

If you want to see the future debate over human enhancement, look no further than today's sports. The modern athlete is a highly enhanced creature. Whatever physiological edge you can get may provide the razor-thin margin for victory in contemporary sports. And with more ways of modifying the body come more restrictions... and innovations to get around the restrictions.

Athletes may very well be leading the rest of society into the debate about who, how, and why people will be allowed — or even required — to enhance their bodies.

Elite players get it all: performance enhancing drugs, surgeries, gadgetry, specialized equipment, even mathematical analysis to help them perform their desired tasks. They are monitored and modeled, tested and retested, sorted and classified. The modern elite player is an isolated cyborgian construct with barely room for a life and identity away from their sport.

Current attitudes towards enhancements vary wildly. Some enhancements are considered the price you pay to get in the game; others, the worst type of cheating. Certain dangerous acts are considered wrong while others are considered honorable. Some seem arcane while others could be useful to anyone and everyone. These attitudes tend to polarize — a new injectable hormone will quickly become anathema, but seeking multiple

LASIK eye surgeries to get better than 20/20 vision is a professional responsibility.

Form matters at least as much as outcome. Take the case of Erythropoietin, or EPO. You make EPO to regulate the number of red blood cells you have, and therefore how readily you can get oxygen to your muscles. Injections of synthetic Erythropoietin to boost performance are a major no-no in sports. It's considered blood doping. But athletes can produce EPO another way: by sleeping in a hypobaric chamber. This reduces oxygen and air pressure to what it would be somewhere 10,000-15,000 feet above sea level. The body responds by producing its own EPO — and lots of it — to get as much oxygen to the sleeping muscles as it can in the deprived environment. After a few weeks in one of these chambers, training in the thick O2 bath at sea level is a breeze. And sleeping in a hypobaric chamber would not be considered cheating any more than pitching a tent halfway up Everest.

Another instructive example is Tommy John surgery, an operation that replaces the ligament in the elbow that tends to suffer most in baseball pitchers. This surgery lets them pitch harder for longer, and despite being a major surgical modification, it isn't viewed negatively. On the other hand, strengthening the arms by supplementing with a combination of testosterone and weight training is prohibited.

What makes a hypobaric chamber OK, but an injection a firing offense? Because we said so.

This may seem hypocritical, but it isn't. After all, the rules of sports are arbitrary. Why shouldn't you use your hands in soccer? Because then it's not soccer. What makes a hypobaric chamber OK, but an injection a firing offense? Because we said so. After we invented agriculture, the bow, or perhaps mountaintop mining equipment, human athletics became a cultural pastime rather than a vital function. No matter how much you love your local sports team, the stakes aren't what they once were. You will not be starved for protein through the long winter if

Barry Bonds isn't hitting like he used to. Thusly, we can pick the rules we like. They don't have to be consistent with anything in the real world.

This is why applying the debate about sports enhancements to the rest of the world can be dangerous. When we're deciding if we should give Modafinil to pilots or Ritalin to grad students, we're making life and death choices about what our future will look like. The questions that arise around sports enhancement — questions about the player's quality of life, autonomy and freedom, or questions around gauging acceptable risk — can help to inform a wider debate on enhancement, as long as we keep those aspects related to arbitrary rules back where they belong — in pastimes.

Botox Parties, Michael Jackson, and the Disillusioned Transhumanist

In Conversation with Christopher Dewdney, culture theorist and author of Last Flesh: Life in the Transhuman Era

http://hplusmagazine.com/articles/enhanced/boto
x-parties-michael-jackson-and-disillusioned-
transhumanist

R.U. Sirius

H+: Michael Jackson seems to reflect various trans-mutant themes.

Michael Dewdney: For me, Michael Jackson represents a sort of pioneer of self-transformation. Aside from whatever questionable personal motives are impelling him, he is using cosmetic surgery to achieve a look that is definitely transhuman. He has taken us by proxy to the frontier of what is currently possible with cosmetic surgery and he has even escaped the constraints of race by lightening his skin color. This last aspect is perhaps the most controversial and disconcerting, but the freedom to choose all your "inherited" features, both familial and racial, will probably become an intrinsic part of the transhuman era.

H+: He reflects, although perhaps not fully consciously, a pursuit of otherness, alienation, and mutation that runs through many contrasting subcultures from psychedelicists to goths to UFO nuts, to early transhumanists, SF fanatics, ad infinitum. And now middle-aged, middle-class ladies have parties to shoot up Botox. Does the mainstream culture show signs of understanding itself as evolving into a mutant breed and do those who need to be different or

avant garde have any new avenues opening up to keep them ahead of the hoi polloi?

CD: The corollary to the Botox craze is the predicament of disillusionment, nay, misanthropism, that I have found myself immersed in the last couple of years. Perhaps the real ground of my disillusionment is my hard-lost benevolence. I'm an optimist; I like people. Yet when I asked a lot of "average" people — people who weren't part of my circle — what they would do with the kind of self-transformative power that may perhaps be ours to wield, I was increasingly appalled. The jocks I talked to wanted to be bigger and stronger so they could beat the shit out of everybody else; the women wanted to morph into their ideal role models. I began to realize that what most people wanted was conformity; their "ideals" would turn us into a world of underachieving Nicole Kidmans and eight-foot Brad Pitts, identical cutouts with no individualism.

My previous rather naive notion that biotechnology would free us from the tyranny of "normalcy," that we could become anything we wanted, morph ourselves into elongated, blue-skinned, orange-haired, sixteen-fingered geniuses or perhaps flying ribbons of sensual bliss that performed acrobatic choreographies above the sunset, was a very utopian and, as it turns out, unpopular dream. Individuality or creative improvisation is the last thing most people want. So Botox is really a dreadful symptom of a new, radical mundanity enabled by biotechnology. And that's disillusioning.

This is Your Brain on Neurotechnology

http://hplusmagazine.com/articles/neuro/your-brain-neurotechnology

Surfdaddy Orca

Zack Lynch is author of *The Neuro Revolution: How Brain Science Is Changing Our World* (St. Martin's Press, July 2009). Neurotechnology is the emerging science of brain imaging and other new tools for both understanding and influencing our brains.

This well-received and well-written book was conceived as a work of popular science "to broaden the conversation" on what Lynch characterizes as the coming neurosociety. Lynch looks at how neurotechnology will impact the financial markets, law enforcement, politics, advertising and marketing, artistic expression, warfare, and even the nature of human spirituality.

The book has received accolades in the mainstream press (including Jane Pauly, at NBC) and from tech figures like Vint Cerf at Google.

Lynch is the founder and executive director of the Neurotechnology Industry Organization (NIO) and co-founder of NeuroInsights. He serves on the advisory boards of the McGovern Institute for Brain Research at MIT, the Center for Neuroeconomic Studies, Science Progress, and SocialText, a social software company.

He earned an M.A in economic geography, and a double B.S. in evolutionary biology and environmental science with high honors from UCLA. His master's thesis examined how the Internet transforms communications and commerce.

You can follow Zack on Twitter at @neurorev

H+: You characterize the Neuro Revolution as the next revolution after the agricultural, industrial, and information revolutions. Others have characterized the Nanotechnology Revolution (for example, the ability to assemble goods at the molecular level) as such a paradigm-shattering period. Do you see a relationship between these two upcoming "revolutions?"

Zack Lunch: Nanotechnology is an enabling technology that will fundamentally drive progress in the neurotech sector. What makes this fundamentally unique, and why the neurotechnology revolution is so profoundly important, is that we are directing our informational and nano technologies at an entirely new domain of human progress: tools for the human mind.

We've spent human history — the past several thousand years as I said in the book — developing tools to improve our physical world. Now we are focused on developing tools that will take our wisdom, knowledge, and capital to develop tools that will improve our inner domain. Nanotechnology will be used to improve the efficiency and effectiveness of drugs, devices, diagnostics, and brain imaging technologies.

H+: You describe a number of emerging neurotechnologies in your book, fMRI being somewhat the granddaddy of the Neuro Revolution. Where should the smart investor be watching for the next fMRI?

ZL: Physics and biochemistry labs. The latest trend in imaging is combined systems –- fMRI and a whole host of other imaging technologies. One of the issues with fMRI is that it's not very good at temporal resolution. What we're trying to do is marry multiple types of imaging technologies to get more refined spatial and temporal resolution in our imaging systems. GE, Philips, and Siemens are developing these combined systems.

H+: Neurotechnology seems like it's an emerging market.

ZL: It's actually a relatively mature market if you consider first generation neurotechnologies. Last year, companies involved in neurotechnology generated about 140 billion dollars in revenue. This includes drugs, medical and neurological devices, and diagnostics for neurological diseases, psychiatric illness, and nervous system injuries. One of the hallmark characteristics of each technological revolution is that when a technology is developed for one purpose — let's say for the purpose of creating treatments for brain or nervous system illnesses — you then begin to see it in a wide variety of different endeavors far beyond it's original intended use.

Who would have thought 10 years ago that we would be using imaging technologies to improve the effectiveness of marketing and advertising? Who would have thought that we would be on the cusp of developing truth detection technologies? Who would have thought that these technologies would be used to understand and perhaps help traders improve their profitability?

What we're seeing across law enforcement, the arts, marketing, entertainment, and warfare is what is means to be human. These technologies are penetrating a wide variety of different endeavors across human society. That — in and of itself — highlights the fact that we are witnessing the very early stages of a Neuro Revolution.

H+: Supercomputers are now faster at leveraging trading positions than humans (this is creating a quite a controversy on Wall Street). What role do you see for human neurofinance and neuroeconomics in the financial markets as artificial intelligence continues to gain more sophistication?

ZL: The technology of each previous revolution is required for the succeeding revolution. We couldn't have had the industrial revolution without the agricultural revolution, because we wouldn't have had the specialization of labor that was required for humans to have the wealth and time to be able to develop industrial technologies. We couldn't have had information technologies prior to industrial technology. In the same way, we couldn't have had neurotechnology without the development of information technology – and without its continued development. These are enabling technologies that will continue to develop, and that will support the evolution of more sophisticated neurotechnologies.

If we're talking about specific technologies that will be available to financial traders, one will be neurosoftware applications that will help retrain the brain of financial traders to reduce the human tendency to overestimate. That will require a quite sophisticated understanding of the human neurobiology of decision making. That — in and of itself — will require computational models that are just beginning to be worked out.

H+: So you envision humans still being in the loop as decision makers rather than supercomputers making sophisticated trading decisions?

ZL: Right. When things like trading become completely automated over time it's because they're based on models. However, the models don't capture all the elements of our world. And part of what drives financial markets — which is what I try to get across in the chapter "Finance with Feelings" — is human emotions.

So theoretically, if you have a financial market where there are no human actors, then computers will be making the decisions. In reality, however, human emotions will sway the market. We need to figure out how to work with those emotions to understand how they actually influence the financial markets, and then leverage them. Emotions are very sophisticated computational algorithms that have been developed through hundreds of millions of years of evolution. They allow us to cut through a tremendous amount of information and give us instantaneous feedback on what we should be doing.

H+: You compare the Neuro Revolution to Copernicus's heliocentric notion of the universe as a game changer that will "bring about new ideas of human spirituality that will forever reshape our understanding of humanity's role and place in the universe." This is akin to what Darwin did in the 19th Century with his theory of evolution. What roadblocks do you see to the societal acceptance of the science of something so fundamental as religious belief?

ZL: In the chapter on neurotheology, I try to tackle the question "Where is God?"

Copernicus's heliocentric notion of the universe and Darwin's theory of evolution transformed how humans look at themselves in the cosmos, as well as here on Earth. I believe that advances in brain imaging technology, in addition to neurostimulation technology, can potentially help reveal a new neurospiritual tradition in the coming decades.

Now, will this come without protest? No. Will this be readily adopted? No. But given the history of how breakthrough technologies have changed our perceptions of our place in the universe, I believe that using neurotechnology and diving deeper into the human brain, we will come to understand the neurobiology of spiritual experiences. This isn't to say that God exists within our brains. I'm not saying one way or another because I don't know. But given history, I'm putting the odds, from my perspective, that we're going to trip over and discover something that will yet again change how we look at ourselves and our place in the universe.

Through neurotechnology we can possibly accelerate peoples' senses of themselves and their relationship to their higher being.

H+: Sex drove the development of both the videotape markets and the Internet as a commercial entity. Your book only has one specific mention of sex, and yet the application of neuroscience and neurotechnology has great potential to enhance human sexuality. What role do you

envision sex playing in the evolution and marketing of the Neuro Revolution?

ZL: (pause) Isn't it sort of obvious? *(laughter)* It's not necessarily a major driving force, but sex is an extraordinarily important component of human behavior – if not *the* fundamental component, along with eating – and neurotechnologies will be developed (and they are being developed) to treat people that can't fully appreciate and experience sex. These same technologies will be used by others to improve their sexual experiences. And this doesn't just include drugs, it includes neurostimulation devices.

H+: You describe both oxytocin (important to human bonding) and dopamine as "emoticeuticals" that will likely have a big impact on a future neurosociety. Will these become FDA regulated drugs like the antidepressants of today?

ZL: The FDA is already involved. But the reality is that we live in a highly complex global economy. If an individual group or a company can develop safe and effective neurotechnology outside the United States, and it then becomes popular and it's used and creates a competitive advantage for an individual or a company or an entire country, then those technologies will seep back into other countries where perhaps they haven't been legalized.

H+: You also mention the example of Adderall and Ritalin as cognitive performance enabling drugs that are being used at the universities. Such drugs, and others like them, have the potential to make students more and more competitive. Do you think that such drugs and upcoming related technology, if used at the national level, could result in something like a neurotechnology arms race?

ZL: Like any new set of tools, the set of emerging neurotechnology tools can be used for both good and bad purposes. Today there are college students, Wall Street financiers, software programmers, and even poker players who are using cognitive-enabling drugs to improve their competitive performance.

There is a whole host of ethical, legal, and societal issues with taking drugs generally developed to treat an illness and then using them to help normal humans. There are issues of safety. There are issues of fairness. There are issues of health. And there are issues of coercion. And the reality is that this is just the tip of the iceberg. For example, there are over 100 compounds in clinical development right now focused on treating some form of memory loss. And we expect a small handful of these over the next decade to improve memory in normal humans. So you can imagine the inherent coercive force that will emerge as those treatments become developed. Imagine a 65-year-old programmer living in San Francisco and she's competing with a 25-year-old in Mumbai, India. Neither one knows whether the other is using one of these cognitive-enabling drugs.

And it's not just drugs; there are neurodevices in development that will be able to improve memory and speed learning. What we're going to see is what I call "neuro competition." This is the next form of competition that individuals and businesses and nations will adapt to gain competitive advantage –- except this will be a neuro advantage. Just as companies today compete for a competitive advantage in information technology –- whether it's the latest social software, the latest IT backbone, the latest servers, or the latest customer relationship management systems –- they will use neurotechnologies to improve their competitive positioning.

The new neurotechnologies will come in multiple forms. They will come not just as drugs to improve one's competitive performance, or emotional performance, or physical stamina, but they will also come in emotion-sensing technologies... and one of the really hard questions moving forward is: where does this all go? This was a major reason for writing the book — to begin to spark a broad public dialog around the societal implications of where this technology might go, and how might we begin to have a conversation around what regulatory options that we might want to start discussing.

H+: Do you see a resurgence of research into substances like LSD as part of the Neuro Revolution?

ZL: There will be continued research into psychoactive substances, although probably not here in the United States. We've got to take into consideration the fact that neuroscientific research is just beginning to emerge in a relatively strong fashion in other countries. It's very sophisticated science –- it's the cutting edge of science. So, while the U.S. and Europe have been leading this basic research in previous decades, emerging companies are coming up from developing nations like China, India, and Brazil where they are starting to develop their own relatively strong neuroscience programs.

H+: You close the book with the idea that a "perception shift" will result from the widespread adoption and application of neurotechnology. What exactly is the role for your Neurotechnology Industry Organization (NIO) in making this come about? And putting on your futurist hat, how quickly do you see this shift occurring?

ZL: NIO is specifically focused — at this point in history — on accelerating the development of treatments for the two billion people worldwide that currently suffer from a neurological disease, psychiatric illness, or nervous system injury. We're trying to drive this forward by working with the U.S. Congress on a national neurotechnology initiative. This will be a new billion dollar Federal R&D program aimed at accelerating transitional neuroscience. A bill was introduced into Congress in March of this year. It would supply 40 million dollars to fund research into the ethical, legal, and societal implications of advancing neuroscience. And when does the perception shift happen? Most likely the 2030s.

Optogenetics:
The Edge of Neural Control

http://hplusmagazine.com/articles/enhanced/enha nced-optogenetics

Quinn Norton

Brain control has always proven tricky, particularly when it comes to the brain trying to control itself. We have many indirect methods — drugs, meditation, education, travel, etc. — but people have always wanted quick and reliable control of their brain states. And what that actually means is that they want to change an area of the brain. Switching the drives and mental states we need on and off would be considerably less frustrating than the transitioning struggles nature has given us. And so we are entering the era of a new set of technologies for direct neural control.

The best current technology combines psychosurgery and implantation. Right now, hard-to-treat disorders can get a difficult direct neural treatment called Deep Brain Stimulation, or DBs. DBs is like a pacemaker for the brain. An electrode is snaked down to the area associated with the disorder being treated and left in place. After the surgery has healed, the implant pulses current at a frequency that either activates or quiets the area responsible for the condition. Affecting cells further from the electrode means passing more current through nearby cells. DBs is by far the most precise clinical procedure for controlling areas of the brain, but it's still disappointingly nonspecific. Since DBs involves brain surgery, it's generally a treatment of last resort, but it's shown good results for previously untreatable cases of Parkinson's, chronic pain, and depression. Electrode implantation is an extreme measure, not likely to be widely used.

Dr. Karl Deisseroth of Stanford University can go one better. He's developed a technique called optogenetics that combines genetic engineering, lasers, neurology and surgery to create a direct control mechanism. Optogenetics uses a brain cell switch with two genetic parts. The first is a gene taken from an algae that activates the cell in the presence of blue light in order to turn towards the light and photosynthesis. In a neuron, that activation fires the cell. The second is from an archaeon, a salt-based extremophile, which responds to yellow light by pumping chlorideions. In a brain cell, that means not firing at all.

To get the genes in place, Deisseroth's team opens up the skull and uses a pipette to apply a nonreproducing adenovirus to the desired brain area. The virus is genetically configured to inject both genes into a single cell type. The single cell will take both genes. After the "light switch" genes are in place, those brain cells are now light sensitive and a 50 micrometer fiber optic cable is fed to the area. In this way, they can target very specific deep brain structures, areas currently too deep and fragile for most psychosurgery. Once the researcher attaches the other end of the cable to a laser, he or she has absolute and flawless control over that group of neurons: blue light on, yellow light off.

Dr. Deisseroth is a psychiatrist as well as bioengineer, and he envisions using optogenetics in place of DBs's not-so-deep cousin, Vagus Nerve Stimulation. Much like DBs, VNs uses an electrode to treat depression and epilepsy but targets where the vagus nerve passes through the neck rather than deep in the brain. It can still cause problems in many patients — sleep apnea, throat pain, coughing, and voice changes are the main complaints. Deisseroth believes optogenetics might be a way of reducing the side effects of VNs by targeting the treatments, rather than just shocking the neck region.

All this points to easier and more effective neural control. We're still far from knowing which cells do what, and further from orchestrating treatments and enhancements for specific conditions. But for the first time we can map and build useful handles on the very things that make us ourselves.

Cognitive Commodities in the Neuro Marketplace

http://hplusmagazine.com/articles/neuro/cogniti
ve-commodities-neuro-marketplace

James Kent

The science of cognitive enhancement is evolving, which means the business of cognitive enhancement is evolving. Supplying cognitive enhancement to the masses can be viewed through the lens of any commodities marketplace. Human experience is already commoditized through drugs that pack mood and performance into portable units — pills or doses — that can be easily traded and consumed, and the drug market is one of the biggest on the planet. The same can be said for audio and visual experience. The platforms and hardware for trading audiovisual experience — TVs, computers, media players, telecomm, cell phones, software — are huge markets with influence over every facet of our lives. The media and drug markets are built upon the ideal of commoditizing consumer moods and experiences. The cognitive enhancement industry is now poised to undergo a similar market revolution.

The cognitive revolution has already begun, as concepts of enhancement move from counterculture and science fiction into mainstream media. Within the last year, the mainstream press has embraced off-label use of Adderall and similar pills as cognitive enhancers for students seeking to better their grades. Soon there will be research to confirm if students using off-label pharmaceuticals get better grades than their peers. The fact that Teva Pharmaceuticals is the corporate supplier of Adderall is rarely mentioned, nor is the fact that these "enhancement" drugs are all copyrighted blends of amphetamines and stimulants marketed to fidgety children. A similar mainstream embrace of students using methamphetamine or cocaine to get better grades will never be seen, because it's in the interest of the media to drive the market for regulated cognitive enhancers and beat the drum against unregulated generic alternatives. All forms of cognitive enhancement — whether a drug or a technology — will face a similar inherent media bias.

Anyone wanting to get into the business of selling moods, memories, and cognitive solutions to the public must first have the interest of the media to help shape market demand. For instance, the same neurostim device that uses electric impulses from a brain implant to treat people with Parkinson's Disease can be tweaked by a few millimeters and pulse rates to make cocaine addicts feel like they are high all the time. Neurostim isn't a cheap commodity yet, but in the future it could be. The "off label" demand for designer neurostim does not exist today, but if the implant procedure was automated and the price was reduced, it could be a very marketable alternative to long-term drug therapy. Cheap neurostim would then fuel an off-label market for cosmetic and personal use with subsidiary markets for designer software upgrades, patches, and applets to customize functionality. But first there needs to be consumer demand for the product, and that has yet to materialize.

The neurostim device that uses electric impulses to treat people with Parkinson's Disease can be used to make cocaine addicts feel high all the time.

The cognitive enhancement revolution may ultimately fail. Comparisons can be made to the Virtual Reality market, which promised a bold age of cyber-living but was encumbered with wonky gear and appealed only to a small number of consumers. Most people prefer watching a very large TV to being goggled into VR — the novelty of a platform doesn't change human preference. VR was clunky, disorienting, and it gave people headaches, motion sickness, and vision problems. Pills with worse side effects are sold by huge corporations, but ultimately VR had no real mass-market application other than coolness. The lesson here is that the success of the platform does not depend on the coolness factor, it depends on consumer demand once the technology becomes affordable. Will the average consumer embrace being implanted, or even crave non-invasive tinkering with memory and intelligence? Modern consumers have embraced taking whatever pill or procedure their doctors recommend, so all perspective next-gen neurotech should take a page from Big Pharma's playbook and pressure MDs to prescribe invasive cognitive solutions to patients for cosmetic and off-label purposes (and pressure insurance companies to cover the costs). Cosmetic therapeutic applications are the doorway to the mainstream consumer market. On the bleeding edge of this field, scientists are already doing research on neurostim to treat depression and sexual dysfunction. (See Resources)

Neural implants and neurostim, like any form of cognitive enhancement, face some challenges with regard to public opinion. The implant procedure is delicate and expensive and could have some unforeseen effects like improper healing or infection. The same can be said of cosmetic surgery or implanting a pacemaker, and the public has adopted those procedures. There are recurring problems with implant interface, hardware, batteries, and security, but the same can be said of iPhones and the public has adopted those. Mix the glamour of surgical self-improvement with the geekiness of high-tech gadget fetishism and you have a niche cosmetic neurostim market waiting to be tapped. The hardware for the neurostim platform is ultimately cheap and automating the procedure is feasible. The applications could enhance memory, intelligence, and mind-to-mind communication. Automating the neural surgery is not impossible — it just takes research grant money and investors. This may seem like science fiction, but in twenty years it may be considered essential consumer technology. It all depends on how the market plays out.

James Kent is the former publisher of Psychedelic Illuminations *and* Trip Magazine. *He currently edits* DoseNation.com[7], *a drug blog featuring news, humor and commentary.*

---

[7] http://dosenation.com/

Will We Eventually Upload Our Minds?

Bruce Katz Interview

http://hplusmagazine.com/articles/neuro/will-we-eventually-upload-our-minds

Surfdaddy Orca

Bruce Katz received his Ph.D. in artificial intelligence from University of Illinois. He is a frequent lecturer in artificial intelligence at the University of Sussex in the U.K and serves as adjunct professor in of Computer Engineering at Drexel University in Philadelphia. Dr. Katz is the accomplished author of Neuroengineering the Future, *Digital Design*, as well as many prestigious journal articles.

Katz believes we are on the cusp of a broad neuro-revolution, one that will radically reshape our views of perception, cognition, emotion and even personal identity. Neuroengineering is rapidly advancing from perceptual aids such as cochlear implants to devices that will enhance and speed up thought. Ultimately, he says, this may free the mind from its bound state in the body to a platform independent existence.

H+: What trends do you see in cognitive enhancement modalities and therapies (drugs, supplements, music, meditation, entrainment, AI and so forth)?

BRUCE KATZ: There are two primary types of cognitive enhancement — enhancement of intelligence and enhancement of creative faculties. Even though creativity is often considered a quasi-mystical process, it may surprise some that we are actually closer to enhancing this aspect of cognition than pure intelligence.

The reason is that intelligence is an unwieldy collection of processes, and creativity is more akin to a state, so it may very well be possible to produce higher levels of creative insight *for a fixed level of intelligence* before we are able to make people smarter in general.

There appear to be three main neurophysiological ingredients that influence the creative process These are 1) relatively low levels of cortical arousal; 2) a relatively flat associative gradient; 3) a judicious amount of noise in the cognitive system. *[Editor's note: A person with a high associative gradient is able to make a few common associations with a stimulus word such as "flight," whereas those with a flat gradient are able to make many associations with the stimulus word. Creative people have been found to have fairly flat gradients, and uncreative people have much steeper gradients.]*

All three ingredients conspire to encourage the conditions whereby cognition runs outside of its normal attractors, and produces new and potentially valuable insights.

Solving compound remote associate (CRA) problems illustrates how these factors work. In a CRA problem, the task is to find a word that is related to three items. For example, given "fountain", "baking", and "pop" the solution would be "soda."

The reason CRA problems are difficult, and why creative insight helps, is that the mind tends to fixate on the stronger associates of the priming words (for example, "music" for "pop"), which in turn inhibits the desired solution.

What are the implications of this for artificially enhancing insight? First, any technique that quiets the mind is likely to have beneficial effects. These include traditional meditative techniques, but possibly also more brute-force technologies such as transcranial magnetic stimulation (TMS). Low frequency pulses (below 1Hz) enable inhibitory processes, and TMS applied in this manner to the frontal cortices could produce the desired result.

Second, the inhibition of the more literal and less associative left hemisphere through similar means could also produce good results. In fact, EEG studies of people solving CRA problems with insight have shown an increase in gamma activity (possibly indicative of conceptual binding activity) in the right but not the left hemisphere just prior to solution.

Finally, the application of noise to the brain, either non-invasively, through TMS, or eventually through direct stimulation may encourage it to be more "playful" and to escape its normal ruts.

In the not too distant future, we may not have to rely on nature to produce the one-in-a-million combination [of a high IQ and creative insight], and be able to produce it at will on many if not all neural substrates.

H+: What are some of the issues (legal, societal, ethical) that you anticipate for such technology?

BK: My own opinion is that — except in the case of minors — we must let an informed public make their own choices. Any government-mandated set of rules will be imperfect, and in any case will deviate from the needs and desires of its individual citizens.

What we in the neuroengineering community should be pushing for is a comprehensive *freedom of thought* initiative, ideally enshrined as a constitutional amendment rather than as a set of clumsy laws. And we should be doing so sooner rather than later, before individual technologies come online, and before we allow the "tyranny of the majority" to control a right that ought to trump all other rights.

H+: What is your vision for the future of cognitive enhancement and neurotechnology in the next 20 years?

BK: Ultimately, we want to be free of the limitations of the human brain. There are just too many inherent difficulties in its kludgy design — provided by evolution — to make it worthwhile to continue along this path.

As I describe in my book, *Neuroengineering the Future*, these kludges include

*   Short-term memory limitations (typically seven plus or minus 2 items)

*   Significant long-term memory limitations (the brain can only hold about as much as a PC hard disk circa 1990)

*   Strong limitations on processing speed (although the brain is a highly parallel system, each neuron is a very slow processor)

*   Bounds on rationality (we are less than fully impartial processors, sometimes significantly so)

*   Bounds on creativity (most people go through their entire lives without making a significant creative contribution to humanity), and perhaps most significantly...

*   Bounds on the number of concepts that can be entertained in consciousness at once (some estimate that the bottleneck of consciousness restricts us to one plus or minus zero items!)

Ultimately, we want to be free of the limitations of the human brain. There are just too many inherent difficulties in its kludgy design...

The alternative is to free the mind from limitations of the brain by the addition of prosthetic devices and ultimately uploading it into digital form. While it is unlikely either of these (and especially the latter) will occur in the next few decades, this remains the ultimate goal of enhancement. Both processing speed and memory will be the most immediate beneficiaries of such developments, but the truly significant gains will involve the types of processing that will be possible.

Freeing the mind from this limited, albeit remarkable, organ will allow us to manipulate thought directly, and this will produce the most gains in intelligence, creativity, and in achieving harmony with other sentient beings and the universe as a whole.

Transhumanism at Play

http://hplusmagazine.com/articles/enhanced/tran
shumanism-play

Pat Kane

Watch children, or adults, at play. And by "play" I mean the real thing — experimental, messy, reality-shifting and explorative, not the routinized pseudo-work we call "leisure" or "recreation."

For kids, their play-world might easily imbue them with strange transformative powers, or they might equally enchant and animate the objects that surround them. For adults — say a bunch of Google engineers bashing and drilling away on the flats at Burning Man — the point of their play is to simulate, as tech historian Fred Turner says, a "utopia of relationships and technology." In either case, when they are at and in play, humans old and young naturally hypothesize about testing the boundaries of human capacities and faculties.

The great guru of play theory, Brian Sutton-Smith, describes play's role in human evolution as that of "adaptive potentiation." By that he means play as the mimicking, mocking or fantasizing about our situations of survival, within zones of time and space that open up in our daily life. In this way, play helps us to improve our ability to respond to the challenges of living. It's our rehearsal hour for real risks and opportunities.

This is why such a flaky, mutable behavior and phenomenon as play has persisted in the human condition. As complex social organisms living with others who are just as complex as us, we've needed the imaginative and hypothetical space it opens up in our daily lives to cope with the strategies, feints and demands of human sociability. We "potentiate" or die.

So play is our evolved and natural capacity to test limits; suspend conditions of reality; imagine our way out of tight situations. But how does this sit with the transhumanist agenda? Doesn't transhumanism take, as a point of principle, that our evolved nature itself is permanently up for being played with and amended, its limits made malleable and even transcendable?

There is — at the very minimum — a positive and negative spin worth considering in this context. Positively, transhumanist ambition could represent the next level of play's evolutionary development within our human condition. Whatever we have done with our fantasies, our flickering simulations, our imaginatively suffused games, we will be able to do with the raw biomaterial of humanity. We then enter into the world imagined by Scottish SF writer Iain Banks in his space operas describing the civilizational challenges of The Culture. The challenge is: how to live well and ethically in a profoundly post-scarcity society, where we have the ability to "play God" with each others' biology and materiality, as a matter of convivial living, and not just upon or over others.

But there's a prior presumption that ethical behavior will kick in at some stage of advancing evolution. The negative spin is that transhumanism may, in actuality, unleash play from its useful psychological netherworld in our species being. In other words, in our imaginations certain types of risk and experimentation doesn't have too much direct consequence. It's just something that keeps the channels of human responsiveness from getting too rusty or ossified, from succumbing to their inherent limits.

The fear is that if we make our bodies, our intentions and their extensions illimitable and thus fully expressive of the "phantasmagoria" (as in Sutton-Smith's descriptions of the transgressions and horrors that he often observes in the coping play of children) of play — then we could be in real trouble.

Play, as it functions in our sociobiology, has to be amoral/non-moral. That's the underground and liminal job it has do — the job of keeping our "potentiation" open, infusing the constraints of human living with indefatigable optimism and possibility. What beauties — but also what monsters — may be made manifest, with our play-drive connected to the transforming technologies predicted by transhumanists? Could it be, in truth, a Pandora's Box: a toy chest filled with Ray Kurzweil's nano-, bio- and robotechnologies?

In 2004, I wrote a book with the pointed title *The Play Ethic*. The title was partly aimed at addressing the fact that the sheer playfulness of our coming society — our ability to "take reality lightly" in so many domains — compels us to think about ethics at the most basic level. How we decide to act humanely in a field of exponentially growing human possibility was, to me, the most urgent of issues — and was obvious related to much of the transhumanist project.

Yet, as the Italian Marxist Paulo Virno says, "there is no objective investigation of human nature that does not carry with it, like a clandestine traveler, at least the trace of a theory of political institutions." The Puritan work ethic presumes a human nature happiest with duty, routine, and social conformism — a useful credo for industrial capitalism. A protean "play ethic" could easily presume a human nature happiest when improvising, being flexible and responsive, exercising imagination: an equally useful narrative, as we know, for informational capitalism.

Each ethic has its supporting cast in the mind sciences. The work ethic is currently undergoing a new intellectual revival, in the age of Obama and his (paraphrasing) economy "built on rock, not sand," (taken from The Sermon on the Mount) and it is bolstered by a new Chicago school of behavioral economics that claims to identify the new "Homer (Simpson) Economicus" in all of us, and argues for a new paternalism to steer (or "nudge") us towards healthy social and economic outcomes.

But a play ethic also has its grounding in neuroresearch that emphasizes the plasticity of the brain: the deeply-founded creativity that generates our consciousness in the first place. Across the opeds, blogs and book review pages, those who want to found their "theory of political institutions" in the next wave of Third Culture science headlines will always have their opportunities.

Yet transhumanism, it seems to me, transcends these familiar political uses of evolved human nature in the sense that it asks us to squarely face our increasing ability to transform that very nature itself, intentionally and by design. And if play operates as dynamically and unpredictably in our unamended nature as I suggest, we are in a moment where we will have to begin to imagine what kinds of "politics" or "ethics" are possible when play's energies are given the most powerful of chariots to drive.

The debate in the late nineties between the German philosophers Peter Sloterdijk and Jürgen Habermas — Sloterdijk a partial enthusiast for transhumanism, Habermas a resolute opponent — generated much heat in certain intellectual circles, but much light too. But it began to hint at exactly what a "play ethic for transhumanism" might be. In his essay, the "Operable Man", Sloterdijk suggests the kind of living-well-together that a profoundly (and materially) playful society might generate.

Biotechnologies and nootechnologies nurture by their very nature a subject that is refined, cooperative, and prone to playing with itself. This subject shapes itself through intercourse with complex texts and hypercomplex contexts. Domination must advance towards its very end, because in its rawness it makes itself impossible. In the inter-intelligently condensed networld, masters and despoilers have hardly any long term chances of success left, while cooperators, promoters, and enrichers fit into more numerous and more adequate slots.

There may be something a little lost in the translation... but the idea that the conditions of transhumanity may lead to subjects that are "refined, cooperative, and prone to playing with themselves" at least splits the difference between the polarities on offer.

Richard Sennett in his recent book, *The Craftsman,* talks of two Greek myths that dramatize our anxiety about technology. It's either Pandora and her box, unleashing all manner of unstable horrors; or the club-footed Hephaestus, whose diligent labor and craft built the palaces of the Gods.

But what of Proteus, Prometheus or Bacchus — those shape-shifters, firebringers and lovers of sensual conviviality? Is there no place for the energetic, mutable, sociable player in transhumanity? No hope for a livable zone that can assuage the fear that transforming technology generates anarchy and thus demands order?

Sloterdijk may be an optimist, but optimism — a deep species based optimism — fuels the play that lurks in all of our breasts. Whatever transhumanists seek to transform in human nature, they would do well to respect the innate transformativity of play itself.

Gamification: Turning Work Into Play

An Interview with David Helgason, CEO of Unity

http://hplusmagazine.com/articles/art-entertainment/gamification-turning-work-play

Ray Huling

Last month, *h+* covered the work of Professor Byron Reeves, who champions the adaptation of gaming technologies for the workplace. Around the same time, David Helgason of Unity, a company that produces game development tools for the Web, mobile phones, and the Wii announced "The Year of Gamification" on the Unity blog (See Resources). For Helgason, gamification is the application of game technology and game design outside "gamespace" and the acceptance of games in non-gaming sectors. Usually, Helgason's customers use his technology to create games like *Zombieville* for the iPhone. But lately, he's noted an increase in customers using Unity to create employee training programs, among other things. *H+* talked with Helgason to get a sense of the practical consequences of gamification.

H+: You've recently seen numerous non-game uses of Unity, one of which is Quartier Saint-Blaise, a model of Paris that allows people to navigate through proposed urban planning projects. What's the story behind Saint-Blaise and what other non-gaming projects use Unity?

David Helgason: I think it was basically a very high-res data set of Paris, taken from planes. You know... where you fly over a city and continuously photograph it and use some analysis technology to turn it into 3-D with textures and everything. I don't know the amount of the data, but it's massive. If you go on our website and walk around on our Island Demo (see Resources), you get to walk around on this tropical island. It's very lush and

239

beautiful. That piece of terrain is a couple of miles on the side. And those guys, I think, had a thousand of those sectors, a massive data set.

Another example that is really cool is something called the Visible Body (see Resources), which somebody has described as a Google Earth for the body. It's an amazing product, from a company out of New York (Argosy Publishing). They put a very high-res, detailed model of the body into Unity and very good tools to kind of peel off the layers and see different bones and nerves and blood vessels. They're licensing that as a tool for medical professionals.

H+: But it's not just the use of game technology that you see spreading to non-game sectors. The use of game design techniques is an important part of what you call 'gamification.'

DH: I've had a lot of good responses to that point, which is about the use of game design methodologies and making other products kind of game-like in the way you interact with them. Mint has done things in this direction, and somebody commented that there's a tax-planning tool like this. They're competing with TurboTax and building game design into the product. It's funny, because it has to be the most boring field, but I mean that's the point. You can make it slightly challenging and give people little reasons to sort of play these tax tools — beyond, you know, not going to prison!

H+: What elements of game design go into gamifying these products?

DH: Game design can be such a pure interaction. I mean, many games are just interaction. There's very little behind them. You're just in the flow of touching something and it moves. It gives you some pleasure and there's a little bit of frustration or stress and you want to overcome this thing. Not all games are like this, but many are. And that skill set... designing that and understanding it and optimizing it so that it feels really good... Getting it right, where people have this pure pleasure from it can be applied to a lot of things. We can see how powerful this iterative process is. I don't know if you get addicted to games, but I certainly have… And I know a lot of people who have.

The serious guys, the military and some of the really big companies like Unilever, have created training packages for some of their employees — and this is where they're coming from. Not necessarily just the 3-D rendering, the fancy, realistic, virtual world experiences, but also the built-in use of frustration and reward.

Training employees on a large scale, companies have often had this problem: how to standardize and roll out good training programs. So they were doing these experiments that I think were successful.

H+: I know I'd feel better about job training if it felt more like killing zombies, but how do non-gaming businesses react to the introduction of both game technology and actual games to the workplace? Is there resistance to this trend?

DH: I hear from people that it can be very all-over-the-map, from very positive to people not understanding what this is all about. Fear and all that.

I was on a panel a while ago, a virtual worlds forum, with a lot of people selling solutions, working with big enterprise, and they spoke of some resistance… but even on the panel, there was a sense that the resistance was going away or that there was less of it now than two years ago.

H+: In some places, you can even find the use of mass market games in corporate training or education. I know of a gaming lounge in New York that rented time on *Ghost Recon Advanced Warfighter,* a squad-shooter, to companies for team-building exercises.

DH: Yeah. I failed to mention this [in the gamification post], but yeah, just using traditional games for various uses, that's obviously true as well.

They did some very large experiments teaching kids with Sim City and The Sims — just playing the games. But these games are extremely rich in knowledge and structural understanding. You can communicate an understanding of a society and how a society works. It was a research project sponsored by Electronic Arts. They rolled out these games and played them in schools, and someone ran around trying to figure out the kids' retention and how well they could apply this knowledge afterwards. The conclusion was that they taught them really well.

In education, you have these terms. One is what you can remember in a multiple choice test right after you learn, and then how much you remember a week after, a month later, and the third is how well you can apply this knowledge in a completely different area.

It turned out that retention was pretty good, but the application of this knowledge was very strong. I'm not an expert in this, but it makes sense to me. You're not actually reading the rules of the game; you're kind of feeling them and internalizing them. People are pretty good at that, and can pick them up quite quickly, even complex rules.

The Perils of FDS Fun Deficiency Syndrome

http://hplusmagazine.com/articles/neuro/perils-fds-fun-deficiency-syndrome

James Kent

Modern cosmetic pharmacology focuses so heavily on eliminating depression that it entirely misses one essential point: depressed people are suffering from a lack of fun. Nobody ever describes depression as a "Fun Deficiency Syndrome," but lack of fun is clearly the root cause of all depression. It is impossible to be depressed when you are having fun, yet modern therapies for depression seek only to minimize depressive symptoms while doing nothing to maximize the daily intake of fun. This backwards approach to treating fun deficiency syndrome — or FDS — is not only dangerously ineffective, it will be viewed by future generations as one of the greatest failures of medicine.

While depression has been studied under a microscope, science has barely scratched the surface on fun. The scientific study of fun is considered to be a frivolous exercise, and this assumption would be correct because fun is frivolous. The mistake made by science and academia is in underestimating the *value* of fun, treating fun as a non-serious diversion instead of a rational goal worthy of scientific examination. This oversight is unfortunate because fun is arguably the greatest thing a human can have. Everyone likes to have fun... no, we *love* to have fun. When we are having fun we forget ourselves and become one with our actions in moments of pure playful enjoyment. Having fun goes beyond being happy. Happiness implies a baseline level of contentment and good feelings but it does not include the amusement, exhilaration, laughter and joy associated with fun. If depression is the illness of our age, fun is the cure.

The roots of FDS can be traced through human developmental stages. Most people have plenty of fun as children, but the onset of adolescence and high school creates a perfect storm of jaded anxiety that dampens the levels of fun easily found in childhood. The onset of FDS in adolescence leads teenagers to naturally seek extremes of fun behavior to counteract their social anxiety. These extremes include partying, fighting, competitive sports and mating behaviors where risk is maximized to produce the most fun. Most people do not consider this adolescent fun-seeking activity to be a neurologically wired behavior to cope with developmental anxiety and depression, but it obviously is. This fun-seeking stage lasts well into early adulthood when chronic FDS becomes more problematic. By middle age, most people are chronically low on fun and this is when depression becomes most acute. If lack of fun is constant and goes untreated it can lead directly to midlife crisis and, eventually, Grumpy Old Fart Syndrome.

Fun can be scientifically reduced to two distinct variables: risk and reward. It is easy to understand why reward is fun, but risk is the key to maximizing the impact of reward to produce fun. The most extreme examples of this dynamic can be found in compulsive behaviors that can become highly addictive, like sex and gambling. Sex and gambling are both fun and risky, and the higher the risk the more satisfying and more fun the reward. Also, consider horror movies or amusement park rides where a constant level of fear and anxiety is sustained throughout the experience until the resolution brings a safe and satisfying reward. Fun is thus the science of using risk to build tension, and then strategically releasing that tension with a pleasurable reward to maximize enjoyment. Fun is therapeutic because it reduces anxiety and produces neurochemicals that combat depression. Fun is one of nature's best and most powerful medicines. If you could put fun in a pill it would almost certainly be illegal.

The major pharmacological variables of the risk/reward fun dynamic are adrenaline and dopamine, the key catecholamines produced in response to stress. By now we should all be familiar with the manic exhilaration of an adrenaline rush and the self satisfied clarity of a dopamine high. Of all the drugs in the world, amphetamines may be the best at stimulating this specific chemical cocktail. It is no mystery why amphetamines lead to risky behaviors. Risky behaviors are even more rewarding under the influence of amphetamines and thus more fun. One side of the dopamine cycle leads the subject to seek out new and fun activities; the other side stimulates the satisfying feeling of reward in response to new experiences. Increasing the levels of risk in these fun-seeking behaviors increases the adrenaline rush and thus increases the sensual intensity of the reward and emotional impact of the resulting memory. The experience of intense fun is therefore more than a trivial diversion: it is a pivotal psychological landmark in the lifetime of an individual that can create long-term changes in self image, mood, and behavior.

If we follow a simple clinical spectrum for FDS, it can be assumed that the longer individuals go without fun, the more depressed they will become. Chronic lack of fun over time will always result in low self esteem and the inability to enjoy activities that were once fun when they were new but have now become mundane. People suffering from chronic FDS will claim to lack the time or motivation to seek out new activities, and at the extreme end of the disorder, subjects will claim that seeking fun is a complete waste of time. This is a chronic lack of dopamine talking, and the only cure for people with FDS is to force them to go out and have fun. Unfortunately subjects with undiagnosed FDS may actually think they don't deserve to have fun, and that they don't even deserve to have friends, so snapping someone with chronic FDS out of their cycle is not always easy. In extreme cases the only solution may be dancing, a surprise party, or a spontaneous and poorly planned road trip. Bring beer.

People are the final component in fun... Other people. Fun is always more fun when it is shared with other people. This is why partying is an essential human behavior for regulating feelings of self-esteem and social worth. Having fun with other humans in a social setting stimulates serotonin and oxcytocin, two neurochemicals essential to feelings of security and being loved. So if you're feeling depressed and nothing seems to be working, the only solution is to call some friends and go out and have some fun. It is clinically proven to make you feel better.

The Perils of CFSS
(Compulsive Fun-Seeking Syndrome)

On the flipside of FDS, we find people who suffer from Compulsive Fun-Seeking syndrome (CFSS). People with CFSS are commonly referred to as adrenaline junkies, thrill seekers, compulsive risk takers and teenagers. While this syndrome is viewed as valuable by the gambling, prostitution, dope, and extreme sports industries, it should be noted that CFSS is a legitimate pathology with a distinct pharmacological profile. CFSS can be artificially simulated by dopamine agonists, including amphetamines, pot, caffeine and alcohol. More oddly, dopamine agonists used to treat Parkinson's Disease or Restless Leg Syndrome (RLS), which selectively stimulate motor pathways and selectively avoid the reward pathways, can also cause compulsive behaviors such as gambling or financial risktaking As the result of CFSS, we find people who chronically seek risky behaviors in the hopes of finding fun, but who fail to feel any long-term satisfaction from the rewards they receive. This syndrome is also called attention-deficit disorder (ADD), or it may be categorized by particular compulsions or addictions, but in actuality these are all symptoms of an underlying CFSS disorder. People with CFSS will become depressed in the absence of fun faster than people who do not have this syndrome.

The Pursuit of Crappiness

http://hplusmagazine.com/articles/humor/pursuit
-crappiness

Joe Quirk

Did you enjoy reading your little transnerdulist magazine about pleasure, prancing and sustained orgasm? Now you expect a little humor, don't you? Well, guess what? I'm not in the mood.

I was in a yoga class this morning, listening to the soothing sounds of the instructor telling me to lean backwards until I stick my nose up my own ass, and if I feel snapping and popping to just go with the flow. The gentle croon of the flute made me grind my teeth, the trickle of the fake waterfall made me have to urinate, and performing the "twisting willow" maneuver made me have to fart. Seeing a mirror everywhere I looked -- that was the final sadism in this crap my wife makes me do to lower my stress. At six goddamn AM.

If you're not from California, you will need me to explain. Yoga is an ancient spiritual practice designed to clear the body of gas. By contorting your joints into exquisitely painful positions, it helps you appreciate life when you're not contorted into exquisitely painful positions. I'm built like a fire hydrant — and I'm about as flexible, so I distracted myself from my seething rage by contemplating this issue's general theme of human happiness.

Try this test. Read some Kahlil Gibran poetry while having your hemorrhoids removed. See if you are profound enough to appreciate the miracle of your moment-by-moment existence. I was audited last year. To sooth myself, I read *Chicken Soup for the Soul*. Didn't help.

Profound people can see the miracle of each moment. Good for them. What about us shallow people? We pay gurus to teach us how to stop being miserable.

Now I'm listening to yoga twerp tell me and a group of huffing, puffing sweaty women that stress does not come from the "world without" but from the "world within." Interesting theory contortionist boy. My personal theory is that stress is caused by everybody constantly pissing me off, and not shutting the hell up while I'm trying to stick my elbow into my armpit because it's good for my health. Hey ladies, how about instead of the "twisting willow" maneuver, we twist the head off yoga boy? That would relieve some stress. Instead, yoga twerp recommends meditation.

Meditation? Guess what that's about? Acknowledging that our default state of mind is a torment. Why do we have to put *effort* into stopping our thoughts from torturing us? Does my dog need to twist his body into contortions to stop himself from neurotic thinking?No, and it ain't because he's spiritual. It's because his species hasn't evolved our style of bloated frontal lobe. That's the part of the human brain that's in charge of imagining long-term future scenarios and choosing among them. Hominids with a deep sense of well-being didn't pass on as many genes as hominids whose anxiety drove them to worry about their children, stress about the next drought, and complain that their hand axes need to be redesigned to prevent calluses. Thus I inherit a brain specialized for bitching. My meditation guru tells me to "observe" my thoughts and "go with the flow," but you don't go with the flow when you're on shit creek.

Are there any legitimate methods for sustaining happiness in the bitching brain? Nancy Etcoff, Harvard psychologist and author of *Survival of the Prettiest*, typed "happiness" into Amazon and found over 2,000 titles that promise to deliver it through the 7 habits, 9 choices, 10 steps, 12 secrets, and 14,000 thoughts of deliriously happy people. This pisses me off. Suppose I read all 2,000 books. Will I be any happier? If anybody had any clue, why would we need 2,000 books?

Tens of millions of us are clinically nuts. How do I know this? There are 120 million medications for anti-depression at large night now.

What do we need to keep our civilization running? Economists will tell you: gas, oil and illegal drugs. Each represents about 8% of the world trade. That's about the right ratio for my personal economy too. I need to drive, buy crap and medicate myself.

Otherwise, I'd have to shoot some of you. We Americans have one of the highest homicide rates among developed countries. This sounds like a big problem until you realize our suicide rate is higher than homicide rate. Most Americans, when given the opportunity to kill the person they hate say, "Oh, to hell with it. It's easier to kill myself."

It's a species-wide behavior. Each year, some 800,000 people across the planet off themselves.

What's our problem? I'll tell you what our problem is. We humans are smart enough to figure out what a raw deal existence is.One of my favorite beach reads is Schopenhauer, who, as far as I'm concerned, made a pretty airtight argument that life is neither good nor indifferent. Life is evil. Life has exquisitely designed every living thing to hurt and kill other living things.

Think about it. Each of us must kill to feed. Nothing that gets fed upon wants to be killed. You will not survive without participating in this evil. There is no way out but death.

Laughing yet? If you refuse to kill to feed, the universe punishes you with a slow torturous death of starvation. The only thing life demands is that you kill. The only sin life punishes is not killing. The only commandment: Thou shalt not not kill.

As The Schopster put it in his chapter, "On the Vanity and Suffering of Life": *This world is the battle-ground of tormented and agonized beings who continue to exist only by each devouring the other. Therefore, every beast of prey in it is the living grave of thousands of others, and its self-maintenance is a chain of torturing deaths.*

If Satan wanted to create the perfect gladiator arena of evil, it would look exactly like the natural world we are in. Schopenhauer says the Secret of Life is: Hell is all that exists, and it requires evil from me if I am to survive. The only way to escape Hell is death. The only way to escape death is to keep killing. And every one of us is doomed to lose.

Bwa! Ha! Ha! Ha! This humor column is on a roll now! When our species became self-conscious and future-thinking, the first thing we noticed is that life is too horrible to live. But death is too terrifying to embrace. Yet there is no third alternative. So we make up a bunch of lies to distract ourselves from the horrible reality of existence.

Go on. Click on another link. Flick another switch on a screen. Stick an earPod into your skull. Twist yourself into a contortion. Go find a god and pray. Anything to distract yourself from the existential hole at the core of your being that drives your ambitions, the acid of self-consciousness that eats away a gaping cavity of boredom that waits for you in the next moment and will continue in an endless march of moments until you die.

Schopenhauer goes further. He says that if empathy really existed, there would be no enjoyment. After all, suffering is everywhere. If we really cared, we'd be perpetually empathizing. But we don't. Why? Because you can't simultaneously be compassionate and content. The existence of enjoyment proves empathy is a fleeting self-indulgence.

So why do we keep struggling to increase the suffering of other sentient beings in order to survive? The illusion of hope. If I keep chasing my next desire, maybe I will finally catch the carrot of sustained happiness that will not wither as I grasp it. That's the hedonistic treadmill.

The only way to end your participation in an evil universe designed to create suffering is to end your mindless Will to survive. If you were truly a being of pure compassion, you would kill yourself right now.

I don't know if this is because I went off my Celexa this week, but it seems to me that Schopenhauer's gift was that he divested himself of the delusions required to survive with a human brain. You want to talk about a guy who stared into the abyss and did not relent until he had used flawless logical steps to march all the way to the bottom. Reading a chapter of Schopenhauer is like listening to a Nirvana album start to finish.

Was he crazy? Psychologists have established that healthy people radically over-estimate the amount of control they have over their lives. The only people who *accurately* assess how much control they have over experimental situations are the clinically depressed. That's right, depressed people are the most uniquely skilled at accurately predicting their control over outcomes. Re-instilling their delusion of control is called curing them. Damn, did those results ever depress me.

So here we are, a self-conscious self-lacerating species whose perpetual sense of dissatisfaction drove us to develop genetic engineering, mood-altering drugs, biotechnology, nanotechnology, artificial intelligence and plastic surgery. We can change the whole ball game. We're all Michael Jackson now. The general response from critics? "Shouldn't we leave well enough alone, trust in the wisdom of nature?"

We may not have the collective wisdom to tinker with nature. But then again, neither does nature. Wisdom teeth don't speak well for the wisdom of nature. I don't think it was very wise to create inside-out retinas that give us blind spots. The pointless bursting appendix, spine and knees incompletely designed for upright tottering, tubes for breathing and swallowing so close we choke, babies who kill their mothers breaching their thick skulls. And whose idea was it for the urethra to pass through the prostate gland? Nature, that's who.

Bad news. Our brains were designed by natural and sexual selection. Our sublime state of complexity and beauty has resulted from a few billion years of struggle which has had no regard for optimizing human happiness. Happiness, as Arthur "Sunshine" Schopenhauer elucidated, is the carrot on the end of the stick that keeps us moving for nature's ends. Rare fleeting nibbles keep us running on the treadmill. Suppose we could rig the stick so we could munch the carrot?

Satisfaction sustained. Empathy mutual. Trust utter. Love permanent. Energy infinite. Violence extinct. Suffering banished. Everything that nature *isn't*.

Stretch. Blank your mind. Be.

###

www.ingramcontent.com/pod-product-compliance
Lightning Source LLC
Chambersburg PA
CBHW051636170526
45167CB00001B/220